Structural Integrity and Reliability in Electronics

Structural Integrity and Reliability in Electronics

Enhancing Performance in a Lead-Free Environment

by

W.J. PLUMBRIDGE
The Open University,
Milton Keynes, United Kingdom

R.J. MATELA
The Open University,
Milton Keynes, United Kingdom

and

A. WESTWATER
Rohm Electronics (UK) Ltd,
Milton Keynes, United Kingdom

KLUWER ACADEMIC PUBLISHERS
DORDRECHT / BOSTON / LONDON

A C.I.P. Catalogue record for this book is available from the Library of Congress.

ISBN 1-4020-1765-0

Published by Kluwer Academic Publishers,
P.O. Box 17, 3300 AA Dordrecht, The Netherlands.

Sold and distributed in North, Central and South America
by Kluwer Academic Publishers,
101 Philip Drive, Norwell, MA 02061, U.S.A.

In all other countries, sold and distributed
by Kluwer Academic Publishers,
P.O. Box 322, 3300 AH Dordrecht, The Netherlands.

Printed on acid-free paper

Contents

PREFACE

Knowledge itself is soon obsolete; It is a blunt instrument.
*Only by **understanding** can problems be solved and progress achieved.*

Reliability in performance of electronic equipment, in the face of demands for continuing miniaturisation and the anticipated abolition of lead-containing solders, represents a major engineering challenge. The involvement of numerous disciplines; such as electrical, electronic, mechanical, manufacturing, and materials engineering together with physicists and computer specialists, adds to the complexity of the situation. Nevertheless, with electronics being the World's largest industrial sector, the potential rewards to the winners are substantial.

This book aims to provide the ingredients for understanding, together with knowledge of reliability in interconnection technology and of the implementation of lead-free solders. It is strongly contended that such a combination forms the necessary basis for greater structural integrity and enhanced performance

The text is essentially in three parts: The intentions of the Part I component (*The Materials Perspective*, Chapters 1-6) are to present a snapshot of the current, but rapidly changing, global scene and to establish a firm understanding of the fundamentals surrounding interconnection performance. With potential readers possessing a broad spectrum of knowledge and expertise, this is essential. It could be argued that the reason for the limited progress made in this field to date has been due to the difficulties encountered in communicating effectively across the discipline boundaries.

Part 2 (*The Manufacturer's Perspective*, Chapters 7-11) outlines the construction of electronic components and factors involved in the production of printed circuit boards which impinge upon their reliability, describes some of the common modes of component failure, and identifies techniques and equipment required to analyse these failures. Improvements in performance and enhanced design methodologies can be achieved only when the exact processes responsible for failure are known. The transition to a lead-free soldering process and the potential reliability issues encountered are also considered.

Using material from Parts 1 and 2 as underpinning, Part 3 (*The Designer's Perspective*, Chapters 12-18) addresses the major challenges of reliable modelling and life prediction. Here, the focus is upon interconnection failure arising from thermomechanical fatigue – the most frequent cause of service failure – although, where appropriate, other mechanisms are considered. Since structural integrity centres upon mechanical behaviour, considerations of damage arising from purely physical or chemical origins are omitted.

The book aims to convey knowledge and understanding regarding key aspects of the reliability challenge. Intentionally, and due to space constraints, it is selective in the information and strategies it describes. Accordingly, referenced work is illustrative and no attempt is made to be comprehensive. While current practice is presented, additional emphasis is placed upon raising the profile of shortcomings in understanding and the availability of appropriate data. This is particularly relevant to Part III.

It is inevitable that some readers will find certain topics to be elementary and obvious, while others may regard the same material as new and challenging. This is the dilemma in writing across discipline boundaries, but it is an essential task. It is hoped that the book will contribute to producing more 'multilingual' engineers and scientists and, if it succeeds, the benefits will be substantial.

RECOMMENDED READING

The following texts constitute a valuable resource in this field, and to avoid constant repetition, they are cited here:

Engineering Materials – An Introduction to Properties and Applications; M F Ashby and D R H Jones, Pergammon Press.

High Temperature Fatigue: Properties and Prediction, (Ed R P Skelton), Elsevier Applied Science.

Solder Joint Reliability: Theory and Applications (Ed J H Lau), Van Nostrand, Reinhold.

Thermal Stress and Strain in Microelectronics Packaging, (Ed J H Lau), Van Nostrand, Reinhold.

Solder Mechanics – A State of the Art Assessment, (Eds D R Frear, W B Jones and K R Kinsman*l*), The Metals Society (USA).

Design and Reliability of Solders and Soldered Interconnections (Eds R K Mahidhara, D R Frear, S M L Sastry, K L Murty, P Kliaw and W L Winterbottom), The Metals Society (USA).

ACKNOWLEDGEMENTS

The authors wish to gratefully acknowledge the assistance they have received from many sources in preparing this book. In particular, Yoshi Kariya, Shellene Cooper, Colin Gagg, Xian Wei Liu and Martin Rist of the Solder Research Group at the Open University made valuable contributions to the initial formulation of the text and its collation. The skill and dedication of Donna Deacon is much appreciated. Thanks are also due to the many colleagues throughout the world who gave permission for reproduction of aspects of their work. The Department of Design and Innovation is acknowledged for financial support.

Professor Bill Plumbridge is Professor of Materials at the Open University and Director of the Solder Research Group. After graduating and researching in Metallurgy at the University of Manchester, he worked in the Engineering Department at the University of Cambridge. Subsequently, he became a Lecturer and Reader in the Department of Mechanical Engineering at the University of Bristol, before taking the Chair in Materials Engineering at the University of Wollongong, Australia. He returned to the Open University in 1991, when he established the Solder Research Group. He has extensive experience in the mechanical behaviour of engineering materials at elevated temperatures, having worked on low alloy and stainless steels, nickel and titanium alloys before focusing on solders. In total, he has published more than 120 papers, about a quarter of which relate to the mechanical behaviour of solders and the structural integrity of interconnections. He is presently Vice-Chair of COST 531 – a pan European collaborative activity on lead-free solder materials, and Chair of the subcommittee for Structural Integrity in Electronics of the European Structural Integrity Society (ESIS).

Angus Westwater has extensive experience in the electronics industry. He has spent some twenty years in research and development, circuit design, equipment manufacturing and PCB assembly processes. In addition, he has spent a similar period in the electronic component industry, working for high profile international semiconductor and passive component companies. During this time he has held positions such as R & D Manager, Technical Support Manager and, more recently, General Manager LSI Marketing, with special responsibility for the management of a component failure analysis laboratory within his employer's quality assurance organisation. He is well known in the industry and professional body-circles, and receives frequent invitations to make presentations at conferences and technical seminars.

Ray Matela received his PhD in Systems Engineering from the University of Waterloo, Canada, and has been at the Open University for 25 years. He has over fifteen years experience of Finite Element Analysis and thirty years of programming. He was awarded, in conjunction with colleagues at Swansea University, a COMETT grant for 500K ECU to develop Multimedia Distance Learning Material for Finite Element Training of Practising Engineers. He is also a consulting engineering analyst with Stress Elements Ltd in the UK. His wide range of research interests include modelling volcanoes, e.g. dykes and slope stability, and modelling creep and thermomechanical fatigue of solder joints. He has authored some twenty papers, three books and holds several patents.

PART 1

THE MATERIALS PERSPECTIVE

In the first chapter of this Part, the developments over the last decade, or so, that have brought about the transition to lead-free solders in electronics are reviewed. The interplay between technical, economic and political interests serves as a good illustration to those who thought 'engineering' was just 'technical engineering'. In effect, the chapter attempts to set the scene, although in this dynamic environment, the scene is far from set. It is always changing. For this reason, many of the reference sources are in electronic form so that the evolution to lead-free technology can be monitored with an immediacy only associated with that medium.

Subsequent chapters develop the mechanical property – structural integrity theme, with emphasis naturally falling on solder alloys – both lead-containing and lead-free, together with those mechanical characteristics most influential in solder joint performance. An elementary approach is adopted for the benefit of readers new to these fields.

Measurement of mechanical properties plays in important role, as does the difference between behaviour of the solder in its bulk form and as part of a soldered joint. Finally, a more quantitative comparison of the mechanical properties lead-containing and lead-free alloys is drawn, and demonstrates that, at least from the performance viewpoint, an optimism regarding the new alloy systems.

CHAPTER 1

SETTING THE SCENE

1. INTRODUCTION

The Electronics industry is amongst the most dynamic and challenging sectors in existence. Competition between manufacturers is intense but the rewards for success are commensurately large. The target of every producer is to be able to offer the market goods that are superior to those already available and at a lower price than its competitors. In the context of electronic equipment, 'superiority' can be associated with improved performance and greater reliability, while 'price' is self-explanatory.

Within this framework, two specific challenges are facing the electronics community. The first is technical and relates to the continuous miniaturisation of equipment, which has placed increasing demands upon structural integrity and reliability of performance in service. The second challenge is environmental and arises from the demand for toxic lead to be removed from solder alloys used for interconnection. This has engendered a search for new alloys, which in turn has placed increased pressures on processing. In addition, to underpin prediction of performance, there is a requirement for information and understanding of the properties of the new alloy systems.

It is important to appreciate the nature of the challenge. The drive for miniaturisation creates the need for improved design and life prediction – the requirement for new lead-free solders compounds this problem. There is considerable debate regarding the environmental significance of replacing lead from solder alloys, but even if lead were to be eventually retained, the objective of reliable and efficient electronic equipment remains.

The challenges described above are extremely multidisciplinary, involving many branches of engineering. To establish some common ground and potential channels of communication, this chapter first outlines the fundamentals of soldering and briefly describes major events of the last decade which have led to the present situation. Subsequently, a sketch of ongoing research activity is presented together with information regarding sources of current progress in this dynamic field. Finally, the emerging candidates to replace lead-containing solder alloys are identified.

2. SOLDER ALLOYS REQUIREMENTS AND APPLICATIONS

This section considers the role and requirements of a typical electronic grade solder alloy to fulfil its processing and performance functions. To facilitate this appreciation, the basics of the soldering process and manufacturing methods will be outlined.

2.1 Basics of Soldering

Soldering is a method used to produce permanent electrical and mechanical connections between metallic materials, neither of which become molten in the joining process. While there are many different processes used in soldering, they all involve four basic ingredients: base metals, solder, flux and heat.

The *base metal* reacts with the molten *solder* to form an *intermetallic compound (IMC),* which establishes a bond between the two materials. This is the basis of a sound joint. The *heat* supplied must be sufficient to melt the solder (usually 30-40°C above the melting point is employed) but not enough to cause any melting of the metals being joined or damage to the board or components.

The thickness of an IMC layer increases with temperature and soldering time, and since these compounds are brittle, the joint may become embrittled and weak if the IMC layer is too thick. In electronics, the base metal is generally copper (or Alloy 42, a nickel-iron alloy), found on the printed circuit board (PCB) metallic circuitry and component leads or pins. If oxidation products form on either the solder or base metals, the quality of the joint is impaired. Therefore, a *flux* is often applied to the surfaces of the base metals prior to soldering to prevent oxidation during heating and to remove any pre-existing oxide layers.

2.2 Soldering methods

By far the commonest methods of soldering large batches of electronic components are wave soldering and reflow soldering.

2.2.1 Wave Soldering

Wave soldering evolved from the early attempts at soldering by dipping the whole printed circuit board into a pot of solder, or dragging the board across the upper surface of a solder pot. This procedure now involves liquid solder being pumped up through a nozzle and allowing it to fall back into the bath, forming a wave (Figure 1-1). Generally, wave soldering is used for the attachment of through–hole components.

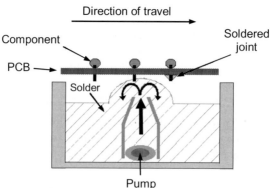

Figure 1-1. The wave soldering process

The PCB, with the components already inserted, travels over the wave via a conveyor system. During this time, solder rises by capillary action up the leads of the plated through-hole components forming the joints. There are several sections of the 'machine' that the board must travel through to enable satisfactory soldering; fluxing, preheating and soldering. Flux is generally applied to the board by spray, and an air knife removes any excess. *Preheating* is necessary to heat up the assembly to the required temperature before the solder is introduced. Preheating reduces the amount of time the liquid solder is in contact with the board, limiting the quantity of intermetallics produced and minimising thermal shock during the process.

2.2.2 Reflow Soldering

Components mounted on the surface of a board require an adhesive to hold them in position if they are to be wave soldered. Figure 1-2 illustrates some common *surface mount components*. In reflow soldering, *solder paste* has the dual function of a source of the solder for the joint and an adhesive for accurate location. Printing solder paste onto the PCB enables an exact quantity of solder alloy in the correct alignment to be supplied for the production of a sound joint. The paste is generally printed onto the copper '*component lands*' on the PCB using a screen printer; components are then placed on top of the solder paste. A stencil is positioned above the board and a squeegee drawn over the stencil forcing paste through the screen apertures and onto the board below (Figure 1-3). The alignment of the board relative to the screen is critical. Incorrect positioning may produce manufacturing faults such as '*solder balling*', '*bridging*' or '*tombstoning*'. These are discussed later.

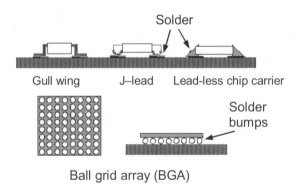

Figure 1-2. Types of components used in surface mount soldering

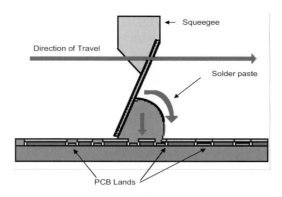

Figure 1-3. Illustration of screen printing onto a PCB

The board and attached components are passed by a conveyor through a multi-zoned oven to preheat the assembly, activate the flux and melt (or reflow) the solder paste. The entire soldering procedure lasts for approximately three minutes. A typical reflow profile is shown in Figure 1-4 although the precise profile depends on the complexity of the board/component arrangement. During the reflow phase, the temperature is maintained at 30-40°C above the melting temperature of the alloy for between 30-90 seconds (preferably under 60 seconds). The appropriate dwell time depends upon the population density of components and the mass of the PCB. If the dwell time is too short, it will result in poor wetting, whereas if it is too long, or the cooling period is extended, the intermetallic layer will be too thick, causing embrittlement of the solder joint. While reflow can take place in air, soldering in an inert atmosphere, such as nitrogen, minimises oxidation inside the process chamber and reduces defects.

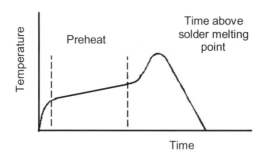

Figure 1-4. Schematic of a typical reflow profile

2.3 Requirements of Solder Alloys

Continuing miniaturisation of electronic equipment and the advent of surface mount technology have put a greater focus on the role of a solder joint, especially with regard to its performance in service. However, it should be recognised that processability (or manufacturability) is a paramount consideration from the commercial perspective. While the principal theme of this text relates to structural integrity and performance, meeting the conditions for effective and efficient processing may have an over-riding impact on the selection of a new alloy. It is important that designers should be sensitive to the limitations imposed at the manufacturing stage.

2.3.1 Processing Requirements

A suitable replacement solder alloy should possess the following attributes.
- a low melting point to conserve energy in production and avoid thermal damage to the board material or to the components;
- the liquid solder should rapidly solidify to allow rapid production and prevent components moving during the soldering process;
- the alloy constituents should have a lower toxicity than lead, and be readily available and economically priced.

Examination of metallic elements with low melting points indicates that cadmium, mercury, thallium and lead may be excluded due to their toxicity. Gallium, bismuth and indium are quite scarce elements, unlikely to be available in sufficient quantities. By far the most promising alloys are based upon tin, which has excellent wetting and spreading properties, is non-toxic and readily available. Alloying tin with a second or third element improves the properties and lowers the melting point.

Alloy systems which are suitable for solders are those that contain a *eutectic* reaction in the phase diagram. The melting point is a minimum for the system, solidification is rapid rather than prolonged with a 'mushy' stage, and the resulting microstructure is fine, conveying good mechanical properties. Hence, eutectic compositions are associated with the lowest costs for heating, minimum thermal damage and the smallest risk of component displacement during the soldering process. These points are best illustrated by consideration of the traditional alloy favoured by the electronics industry, tin-37 weight percent lead (Sn-37Pb). This will provide a good basis for an appreciation of the new generation of lead-free solder alloys.

Lead forms a simple eutectic with tin, and this provides an (almost) ideal alloy that has been in use for over a century. (For other applications it has been in use for thousands of years). A schematic of the tin-lead phase diagram is shown in Figure 1-5. The eutectic composition is given on this diagram at 63% tin and the eutectic temperature as 183°C. (With many alloy systems, there is often some dispute as to the exact composition that transformations, such as the eutectic, occur. It is only when this involves infringing tight patent specifications that this becomes a real problem!). In this diagram, α and β are solid solutions of tin in lead and lead in tin respectively.

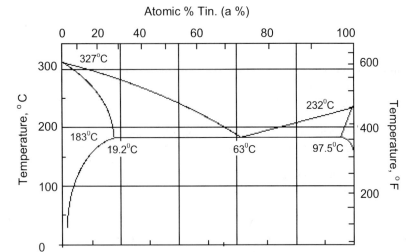

Figure 1-5. The tin-lead phase diagram

A phase (or equilibrium) diagram describes solely events occurring under equilibrium conditions, i.e. during slow cooling. But slow cooling of molten solder does not occur in practice - nor is it desirable. Figure 1-6 demonstrates the significant difference in microstructure between slow and rapidly cooled conditions. In reality, a range of microstructures is likely in the joints on a PCB according to the specific cooling rates that individual joints experience.

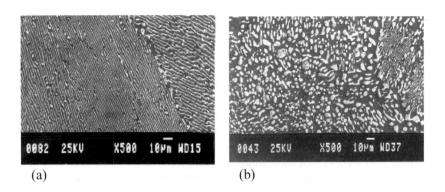

(a)   (b)

Figure 1-6. Effect of cooling rate on the microstructure of a eutectic lead-tin solder. (a) slow cooled; (b) rapidly cooled

The other advantages of using lead are;
- its relatively low cost,
- its ability to reduce the surface tension of tin enabling the solder to flow and spread better (i.e. improving 'wettability/solderability'),

- its limitation on the production of intermetallics by reducing the rate at which substrate materials are dissolved by tin,
- its significant solid solubility in tin which strengthens the tin-rich phase,
- the wealth of information and experience has been accumulated on this alloy.

Taking this into to account, it is not surprising that the discovery of a complete drop-in replacement alloy represents a formidable challenge. To minimise problems encountered in the transition to lead-free technology, manufacturers would prefer the melting point of the replacement alloys to be as close as possible to that of Sn-37Pb.

For elevated temperature applications around 300°C, alloys with higher lead contents (such as Pb-5Sn, Pb-2.5Ag-1Sn) are employed. Little progress has been made in developing lead-free alloys in this domain.

2.3.2 Performance Requirements

The conditions under which electronic assemblies are required to operate vary dramatically with application. Temperature is frequently the biggest variant in service and, depending upon application, can range from –55°C to 180°C. Typical examples include,

Consumer electronics and telecommunications equipment: –20°C to 100°C

Military applications: –55°C to 125°C

Aerospace and automotive applications: –55°C to 180°C

Naturally, these domains are highly application-specific and the trend is for the degree of severity to increase. The *Reliability* of a component or system may be defined as the probability that it will perform its intended function for a specified period of time, under a given operating condition, without failure. Again, acceptable values for reliability are application dependent. The various mechanisms of failure are discussed in subsequent chapters.

3. BACKGROUND EVENTS

This section briefly recounts some of the important events during the last decade that have led to the present situation. It should be emphasised, at the outset, that the circumstances are highly flexible (sometimes reversible!), and that what is described on one day may not be the case shortly afterwards. With that caveat, the technical, environmental, political, commercial and scientific perspectives of the evolution of lead-free soldering and its use in electronics equipment are outlined.

For many years, the most popular general-purpose alloy used in the joining of electronic components to printed circuit boards and in the manufacture of components for die attachment has been the tin-lead alloy, usually at, or near the eutectic composition of Sn-37Pb. It is well known that lead is a toxic metal and can be the cause many serious health issues, particularly in children. It has, in the past, been the target of several environmental campaigns; for example, legislation in the USA in 1986 (*The Safe Drinking Water Act* [1]) was enforced, banning lead-containing solder alloys from use in plumbing. Concerns have also arisen regarding solder use in other industries, in particular, electronics. Lead can contaminate the environment during disposal of 'end of life' electronic products. Discarding equipment either into landfill or by incineration can have a detrimental effect on the environment. When disposed into landfill, lead can be leached from the soldered joints under the action of acid rain. Lead leachant can then enter into ground water and eventually contaminate domestic water supplies.

Alternately, if 'end of life' equipment is disposed of by incineration, airborne particles of lead-containing materials could be released into the atmosphere contaminating surrounding farmland. Lead can be ingested into the body during the consumption of contaminated crops. In response to these concerns, various legislative actions have been initiated to eliminate or reduce lead in electronic equipment – this may be described as the *environmental push*.

3.1 The Drivers of Change

The early driving force towards lead-free electronics, from the environmental perspective, began in the United States during the 1990s. This was the result of the introduction of the *Reid Bill, S.391 – Lead Reduction Act* in 1991 (2) and the *Lead Exposure Reduction Act, S.729* in 1993 (3). However, the electronics industry resisted the intentions of these bills, claiming that manufacturing with lead-free solders was not possible, and the legislative efforts were shelved. There has been little subsequent effort to legislate against lead in solders in the US, although in 2001, the *Environmental Protection Agency* (EPA) proposed a crack down on lead emissions from industrial plants (4). This may in turn impact on the electronics industry.

In Europe, the pressure on the electronics industry for the elimination of lead has been more formidable. The European Commission (EC) has published draft proposals *(Waste Electrical and Electronic Equipment (WEEE)* (5) concerning restrictions on the disposal of hazardous material in landfill from 2006. The *Restriction of Hazardous Substances Directive* also bans lead (along with other hazardous materials) from electronic assemblies from 1st July 2006 although the final details are awaiting confirmation. This will not only impact upon solder alloys, themselves, but also on component finishes, board finishes and flame retardancy issues. These directives have been subject to much debate and, when finally adopted, the entire European Community and suppliers to it will be required to comply. It is estimated that up to 30 per cent of lead in a solder joint originates from sources other than the solder itself - board, component and leads - so there is scope for further definition!. The EC has also passed a Directive on *End of Life Vehicles* (6), which was adopted in September 2000. This targeted recycling and reuse, but has additional clauses affecting the use of hazardous materials. Solders used for automotive applications have a temporary exemption from the ban. The main objectives of the *WEEE Directive* are:
-   improved design of products to avoid generation of waste;
-   manufacturer responsibility for certain phases of waste management;
-   separate collections of electronics waste;
-   appropriate systems established by manufacturers to improve treatment and reuse/recycling of electronics waste.

Legislation in Japan deals with the reclamation and recycling of household electrical appliances. In 1998, the *Japanese Ministry of International Trade and Industry* (MITI) decreed that as from April, 2001, manufacturers would be responsible for the reclamation of all lead used in equipment they produced (7). This and equivalent legislation has proved a powerful driver for manufacturers of electrical goods to move towards completely lead-free products. Similarly, most of the major electronics producers have announced time scales in which lead will be phased out of their products. The majority expect to be producing significant proportions, if not all, lead-free products by 2003.

In contrast to the 'environmental/legislative push' towards lead-free soldering, there are various factors providing a *commercial pull*. This is now proving to be more significant in driving the transition away from conventional soldering materials. The 'classic' example of the environmental pull occurred when sales of a (Matsushita) mini disc player rose from 4 to 14 per cent of market share once a lead-free 'Green Leaf' label was added to the product. Consumers' environmental awareness will continue to grow, and countries wishing to compete with, or export to those already using lead-free solders will have to convert. It has been this commercial factor, rather than the legislative threat that has induced US companies, somewhat belatedly, to embrace the transition to lead-free solders.

In addition to environmental push and commercial pull, a *technical* element is contributing to the transition in electronics assemblies. The requirement of a solder joint is changing. As printed circuit technology advances, the sizes of components, and hence 'end products', are reducing. Figure 1-7 showing the miniaturisation of mobile phones in recent years, is a typical example of this trend.

Figure 1-7. Miniaturisation of mobile phones over the past five years

Surface mount technology (SMT) is replacing the traditional through-hole method of manufacture. This allows attachment of smaller components to both sides of a printed circuit board, facilitating miniaturisation (Figure 1-2). However, it places a greater significance on the quality and reliability of the solder joint, which now has an important role in *structural integrity,* in addition to its usual function as a conductor of heat and electricity. A consequence of this is that the mechanical behaviour of the solder and the joint are becoming increasingly influential in determining overall board performance and reliability. Greater attention will need to be paid to the comparative mechanical properties of the new generation of lead-free alloys and those of lead-tin. The range of mechanical properties that may influence service life is wide but preliminary studies on alternative alloys have provided optimistic indications (8, 9).

3.2 Industry's Response

Several companies, particularly in Japan, are already using lead-free alloys in their production lines. Worldwide, many have announced time scales over which they intend to implement lead-free technologies. However, any detailed overview of industrial activity in this direction would be out of date as soon as it was prepared. Apart from matters of confidentiality, the multinational nature of the major players and the extensive networking and collaboration that occurs render any such review immediately redundant. The reader is referred to the plethora of current information that appears on the Internet.

Perhaps, the most notable overviews were snapshot surveys commissioned by *The Department of Trade and Industry (DTI)* and published in 1999 as *Lead-free Soldering*, (10) and in 2000 as *'Lead-free soldering - One year on'* (11). Discussions over a two-month period with industrialists and researchers were converted into an informative and comprehensive booklet covering Materials, Technology, Research and Development, Legislation and Dissemination. The two reports constitute an excellent introduction to the field.

4. CURRENT LEAD-FREE RESEARCH

Many national and international consortia have carried out research on lead-free soldering. The earlier programmes concentrated on defining possible alloy systems that could replace tin-lead, and their basic properties. The criteria employed included toxicity, melting point and range, cost and availability. It was concluded that tin-based alloys would be the most likely replacement although the concept of a completely 'drop-in' alloy seemed unlikely. The recent programmes concentrate on the major alloys that were identified initially and involve more detailed examination of key properties of these alloys. Until recently, research in the U.S. was limited. However with Europe and Japan actively pursuing the 'lead-free goal', the U.S. industries have resumed research to ensure their global competitiveness.

Table 1.1 indicates a selection of the principal research organisations contributing in various ways to the challenges facing structural integrity and reliability of electronics equipment. It is by no means comprehensive and reflects the personal experience of the authors. However, in conjunction with the List of Information Sources presented in Section 6, it constitutes a useful starting point for more in-depth investigation.

A landmark study into solder alloys to replace lead-tin was the *Brite Euram* funded *'Improved Design life and environmentally aware manufacturing of Electronics Assemblies by Lead-free Soldering' (IDEALS)* programme (12) that involved a consortium of major European companies. The principal outcome was the recommendation of Sn-3.8Ag-0.7Cu (or similar compositions) as the general-purpose replacement alloy, together with the finding that many of the lead-free alternative alloys possessed the necessary processing properties and were stronger and more reliable in thermomechanical fatigue than Sn-37Pb. In effect, this study demonstrated that, contrary to the earlier concerns of American industry, lead-free soldering was a practical proposition with both direct and indirect benefits'.

Table 1-1. A Selection of On-Going Research on Lead-Free Solders and Structural Integrity

Organisation	Country	Activities
The Open University	UK	Mechanical properties of solders and joints, constitutive equations, life prediction.
The University of Greenwich	UK	Modelling solder joint formation and performance.
Loughborough University	UK	Processing effects on joint quality and performance.
National Physical Laboratory	UK	Processing properties, inspection, thermal cycling PCBs, miniature specimen testing, thermodynamic calculations.
International Tin Research Institute (Soldertec)	UK	Information broker, database, advice centre, dissemination and training
The Welding Institute	UK	Solderability, thermal cycling and reliability of PCBs.
Fraunhofer Inst. Berlin	Germany	Mechanical behaviour, microdeformation, thermomechanical testing, life estimation.
University of Vienna	Austria	Mechanical properties, thermal strains, oxidatic EMF measurements, size effects.
Chalmers Univ. of Technology	Switzerland	Thermodynamic assessment, wettability, fatigu life prediction, degradation.
Institute of Production Engineering Research	Switzerland	Fatigue of joints, long term reliability during thermal cycling.
EMPA (Federal Labs for Materials Testing and Researcl	Switzerland	Mechanical properties, NDT, modelling phase diagrams and performance.
Foundry Research Inst.	Poland	Wetting, interface structures and strength.
University of Metz.	France	Nanoparticle and nanograin strengthening in joints.
Fraunhofer Inst. IZM, Berlin	Germany	Constitutive equations, fatigue, mechanical behaviour of joints.
Siemens	Germany	Mechanical properties, creep modelling, reliability, damage.
University of Toronto	Canada	Conductive anodic filaments, fluxes, processin microstructure, thermal fatigue, accelerated testing, modelling.
Michigan State University	USA	Fatigue and thermomechanical fatigue of joints microstructural and dwell effects.
Ames Laboratory	USA	Microstructure – property relationship in joints
University of California	USA	Creep
Shibaura Inst.	Japan	Effects of alloy additions on strength and fatigu behaviour, life prediction, interface structures and reliability.
Ritsumeikan University	Japan	Mechanical properties, life prediction, standarc
Yokohama University	Japan	Finite element analysis and life prediction of joints.

Effective implementation of lead-free technology is a global challenge, and the goal of reliability and efficient design in the face of continuing miniaturisation requires global resources. Extensive collaborations between continents, countries, industry and academia have been established to perform the necessary research and to provide conduits for dissemination of information. Again, these networks or groups are valuable sources of information. Such collaborations include:

EMERnet: (Electronics Manufacturing Engineering Research Network) is a UK-wide electronics manufacturing research network. Its aims include maintaining, promoting and building upon the high quality research in this sector and improving industrial collaboration, dissemination and exploitation.

The SMART Group: (For the advancement of *S*urface *M*ount *A*nd *R*elated *T*echnologies) is a leading European technical trade association representing the electronics manufacturing industry. It promotes the industry through education, training and dissemination.

The Printed Circuit Interconnection Federation (PCIF) is a trade federation representing the UK PCB industry. It aims to enhance members' position in global markets, represent the industry at technical and government levels, provide the resources that enable this representation to be effective and assist in shaping the future of the interconnection industry.

National Centre for Manufacturing Sciences (NCMS) is a non-profit making research consortium of North American corporations. Research work is currently focused upon specific alloy groups along with flux development to improve solderability, reliability and joint appearance.

National Electronics Manufacturing Initiative (NEMI) is a partnership of the North American electronics manufacturing industry. It has been actively involved in recommending alternative alloys through the NEMI Lead-free Assembly group.

Equivalent organisations in Japan, including the *Japanese Institute of Electronics Packaging* and *New Energy and Industrial Technology Development Organisation* (JIEP and NEDO), aim to find standard lead-free solder and soldering conditions.

Two European consortia are promoting the development and implementation of lead-free technology both internally and with equivalent organisations on other continents. Under the EU system for Co-operation in the field of Science and Technical Research, COST 531 (Lead-free Solder Materials) is an extensive programme encompassing theoretical prediction of new alloy systems, evaluation of processing and performance properties, reliability and packaging issues. While attempting to understand and model behaviour of the first generation of lead-free solders, its longer-term goal is the prescription of its successors. ELPHNET, the European Lead-free Network, has broadly similar goals, with greater focus upon implementation, dissemination and co-ordination of activities although there is a central core of common membership.

5. ALTERNATIVE SOLDER ALLOYS

The aim to meet the legislative and technical requirements, described above, for lead-free solder alloys, engendered considerable international effort. Fortunately, the outcomes were broadly similar, i.e. there was no obvious drop-in replacement for eutectic tin-lead, and there was a reasonable consensus as to which alloys might be employed. The majority of possible alternative alloys are based upon one of the following five groups: Tin-silver, tin-copper, tin-silver-copper, tin-silver-bismuth and tin-zinc-bismuth. Table 1.2 shows the compositions and melting points of some of the alternative alloys.

Table 1-2. Solder Compositions and Melting Points ((e) denotes the eutectic composition) (10)

Alloy System	Composition (wt %)	Melting Range (°C)
Sn-Bi	Sn-58Bi	138 (e)
Sn-Bi-Zn	Sn-8Zn-3Bi	189-199
Sn-Ag	Sn-3.5Ag	221 (e)
Sn-Cu	Sn-0.7Cu	227 (e)
Sn-Ag-Bi	Sn-3.5Ag-3Bi	206-213
Sn-Ag-Cu	Sn-3.8Ag-0.7Cu	217
Sn-Ag-Cu-Sb	Sn-2Ag-0.8Cu-0.5Sb	216-222

The characteristics of each of the alloy groups are considered below:

Sn-Cu (Sn-0.7Cu): This is one of the cheapest lead-free alloys available. However, tin-0.7copper has the highest melting temperature (see table 1.2) and would probably incur higher manufacturing costs. In addition, the alloy possesses poor mechanical properties in comparison to the other lead-free candidates (9), and it may be susceptible to the formation of tin-pest (see Chapter 2) when subjected to temperatures below 13°C for prolonged periods (13).

Sn-Ag (Sn-3.5Ag): Tin-3.5 silver is a eutectic alloy with a melting point of 221°C. Its reliability and strength are good, exceeding those of tin-copper and tin-lead, particularly in respect of creep resistance (9).

Sn-Ag-Bi (Sn-3.5Ag-3Bi): This alloy has a lower melting point than other lead-free alternatives, and minor compositional variations melt in the range 200-210°C. The solderability is the best of the replacement alloys (10) but when in contact with Sn-Pb component finishes, this alloy shows a tendency to exhibit fillet lifting or hot tearing during solidification.

Sn-Zn-Bi (Sn-8Zn-3Bi): Athough alloys based on Sn-Zn can be produced with melting temperatures very close to that of the Sn-Pb eutectic, the presence of reactive zinc causes problems such as corrosion, excess drossing and oxidation (10). Some workers consider that these shortcomings could be alleviated by improved fluxing. The mechanical properties of this alloy are broadly similar to those of eutectic lead-tin, with the creep resistance being superior at high stress levels (14).

Sn-Ag-Cu (Sn-3.8Ag-0.7Cu): The tin-silver-copper alloy was developed as an improvement on the basic Sn-Ag alloy. It melts via a ternary eutectic at 217°C, although the actual ternary eutectic composition is still a matter of debate. The Sn-3.8Ag-0.7Cu alloy was recommended for general-purpose use by the IDEALS project (8), and the preferred composition in USA is Sn-3.9Ag-0.6Cu. It has a higher strength and a slightly lower ductility than Sn-Ag and Sn-Cu. The alloy composition falls just outside a Patent (covering any alloy in the range of Sn-3.5-7.7Ag-1.0-4Cu, and using at least 89 percent tin) and so avoids compulsory purchase or royalties. Its mechanical properties, and those of a alternative alloy (Sn-4.0Ag-0.5Cu) designed for this purpose, are similar to those of the patented composition (15). This is another good example of financial considerations influencing technical decisions.

6. USEFUL INFORMATION RESOURCES

Table 1.3 lists a variety of sources of information regarding lead-free technology. Again, no claim for its completeness is made although those sites shown, together with their own recommendations, constitute a formidable resource.

Table 1-3. Resources on Lead-Free Solders

Organisation	Website
Indium Corp.	http://Pb-free.com
IPC dedicated Pb-Free site	http://www.leadfree.org/
International Tin Research Institute Ltd (Soldertec):	http://www.lead-free.org/
Japan Electronic Industry Development Association:	http://www.jeida.or.jp/guide/gaiyou/index-e.htm
DTI Lead-Free Soldering Report.	http://www.npl.co.uk/npl/ei/research/leadfree.html or http://lead-free.org
Ministry of International Trade and Industry:	http://www.miti.go.jp/index-e.html
National Centre for Manufacturing Sciences:	http://www.ncms.org
New Energy and Industrial Technology Development Organisation	http://www.nedo.go.jp/english/index.html
National Electronics Manufacturing Initiative, Lead Free Interconnect Project:	http://www.nemi.org/PbFreePUBLIC/
National Institute of Standards and Technology	http://www.nist.gov
The Open University Solder Research Group	http://materials.open.ac.uk/srg/srg-index.html
Waste from Electrical & Electronic Equipment (WEEE Report)	http://www.lead-free.org/
ELPHNET	http://www.tintechnology.com
COST531	http://www.ap.univie.ac.at/users/www.COST5:

7. CONCLUSIONS

Solders are required to offer mechanical, electrical and thermal continuity to electronics assemblies. To avoid thermal damage to the PCB, or other plastic materials, the temperature range for soldering is narrow. At present, tin-lead eutectic solder satisfies these requirements well, but environmental and commercial pressures demand alloys without lead. Several lead-free alternatives have been identified, and a ternary Sn-Ag-Cu alloy is emerging to be the favourite for general-purpose applications. For low temperature use, Sn-58Bi and Sn-52In are likely candidates, but no substitute is obvious for high temperature applications. Relatively little is known about these emerging alloys and their more complex mechanical behaviour which influences service performance.

Two key factors differentiate traditional lead-tin solder from the new lead-free alloys. First, the higher melting point (30 – 40°C) of the latter imposes a significant penalty in terms of heating and potential thermal damage. Possible benefits include greater stability of microstructure and higher strength. Secondly, the amount of solute in tin is generally substantially different (approximately 40 per cent as compared with a few per cent, at most). This means that the alloys are quite different in microstructure, as will be their response to service conditions. The temptation to make direct comparisons requires careful consideration.

Over the last decade, the climate has changed from a push towards lead-free soldering driven by environmental forces. The goal is unchanged, but the main justification now is a commercial one. The re-entry of US industry means that there is a global race underway to develop lead-free solder alloys, with little or no extra cost in materials or processing, which perform at least as well as the traditional lead-containing solder alloys. An increased share of the electronics market is a valuable prize!

8. REFERENCES

1   Public-Law (99-339) Safe Drinking Water Act Amendments, 1986: United States of America.
2   US Senate and House of Representatives: S391, Lead Exposum Reduction Act, 1991.
3   US Senate and House of Representatives: S729, Lead Exposum Reduction Act, 1993.
4   L Hester, Environmental News, 2001, EPA, Washington DC.
5   Proposal for Directive of the European Parliament and of the Council on Waste Electrical and Electronic Equipment, June 2000.
6   European Commission Directive 2000/53/EC End-of-Life Vehicles. September 2000.
7   Ministry of International Trade and Industry, http:\www.miti.go.jp. 2000.
8   DM Jacobson, and MR Harrison., Brite Euram BE95-1994 IDEALS. GEC J. Research, 1997, 14(2).
9   WJ Plumbridge, CR Gagg, and S Peters, J Electronic Materials, 2001, 30, 1178.
10 B Richards, CL Levogner, CP Hunt, K Nimmo, S Peters and P Cusack, Lead Free Soldering - An Analysis of the Current Status of Lead-free Soldering. 1999: DTI.
11 B Richards and K Nimmo, An Analysis of the Current Status of Lead-free Soldering - one year on., 2000, DTI. p. 1-11.
12 MR Harrison, JH Vincent and HAH Steen, Soldering and Surface Mount Technology, 2001, 13, 21.
13 Y Kariya, CR Gagg and WJ Plumbridge, J of Matls, 2001, 13, 39.
14 I Shohji, CR Gagg and WJ Plumbridge, J Electronic Matls, to be published.
15 Y Kariya and WJ Plumbridge, Seventh Symp. on Microjoining and Assembly Technology in Electronics, MATE 2001, (Yokohama), 2001.

CHAPTER 2

INTRODUCTION TO THE PROPERTIES OF MATERIALS

1. INTRODUCTION

There are four basic classes of materials: *Metals, Ceramics, Polymers and Composites* (which are often mixtures of two of the classes). Somewhat unusually for engineering structures, all types may be found in, or adjacent to, a soldered interconnection. Therefore, an appreciation of the major characteristics of each group is a valuable aid in understanding the behaviour of actual joints which generally determines overall performance.

This chapter will compare and contrast those properties of the four material classes that are influential in service performance. Since a joint may be regarded as a structure comprising materials from various groups, the features of potential concern are any significant *differences* that exist in the property values. Unfortunately, this situation is not uncommon. While structural integrity is primarily a mechanical attribute, much of the architecture on a PCB is dictated by the physical properties of the materials involved, and these are considered initially.

2. PHYSICAL PROPERTIES

Physical properties of materials that are important in both joint design and performance include *electrical conductivity* and *thermal conductivity*. Of the material classes, only metals may be regarded as good conductors of electricity. (The properties of a composite are approximately the average of those of its components, taking into account the proportions involved). The differences in electrical conductivity are enormous - many orders of magnitude (Table 2.1). Since one role for solder material is to provide an electrical conduction path, this is why solders are nearly always metals. An exception to this is the emerging conductive polymer which is a composite containing a polymer base with metallic particles to provide the conduction.

With regard to thermal conductivity, similar characteristics exist; metals are generally very good conductors of heat whereas the other material groups may be described as *insulators*. Again, the differences are substantial (Table 2.2). Therefore, any heat generated in a component will be easily conducted away when in contact with a metal, but thermal efficiency is considerably reduced by the presence of a polymer or ceramic, and that may lead to damaging high temperatures.

The ability of a material to resist high temperatures is principally determined by its *melting point*. In this case, ceramic materials excel and polymers fare badly (Table 2.3). Metals have intermediate melting points, although those used in soldering are quite low for processing and compatibility reasons. For example, the common Sn-37 Pb solder alloy melts at 183°C and the popular lead-free alloys, such as Sn-3.5 Ag and Sn-0.7 Cu, melt at around 220°C.

Table 2-1. Electrical Conductivity of Materials at Room Temperature

Material	Electrical Conductivity $(ohm-m)^{-1}$	Material	Electrical Conductivity $(ohm-m)^{-1}$
Silver	6.8×10^7	Lead	4.9×10^6
Copper	6.0×10^7	Solder (Sn-37Pb)	6.8×10^6
Gold	4.3×10^7	Silicon	4×10^{-4}
Aluminium	3.8×10^7	Alumina	$10^{-10} - 10^{-12}$
Brass	1.6×10^7	Nylon	$10^{-9} - 10^{-12}$
Tin	7.9×10^6	Polyethylene	$10^{-13} - 10^{-17}$
		Polystyrene	$< 10^{-14}$

Table 2-2. Thermal Conductivity of Materials

Material	Thermal Conductivity $(wm^{-1} K^{-1})$	Material	Thermal Conductivity $(wm^{-1} K^{-1})$
Silver	428	Lead	35.3
Copper	398	Solder (Sn-37Pb)	50.2
Gold	315	Alumina	30.1
Aluminium	247	Glass/Ceramic	5.0
Silicon	150	Nylon	0.24
Brass	120	Polyethylene	0.38
Tin	66.8	Polystyrene	0.13

Since components in electronics equipment experience a range of temperatures during service, then the *coefficient of thermal expansion* (CTE) is important, particularly where materials are fixed together, as in the case of solder joints. Table 2.4 indicates that the extent of these differences can be significant. The generation of stress and strain under conditions of changing temperature, when materials with different thermal expansion coefficients are joined, is the principal cause of failure of soldered joints in service. The CTE of the common lead-free alloys is less than that of Sn-37Pb which ameliorates the problem slightly. The capacity of materials to accommodate these stresses and strains is largely determined by their mechanical properties, which are considered in Section 3.

In the absence of lead, the *density* of lead-free solders is reduced. While this results in more joints per unit mass of solder, it has a practical downside. Traditional liquid Sn-37Pb solder is sufficiently dense to allow foreign bodies made from steel, to float on its surface when they accidentally fall into the solder bath. However, they will sink in a bath of lead-free solder alloy, with a risk of serious damage to the bath impeller. Yet another example of a spanner in the works!

Differences in surface tension and optical properties causes lead-free solder joints to be shallower and have a less reflective appearance than joints using Sn-37Pb solder. Inspection methods require modification.

Table 2-3. Melting and Softening (s) Temperatures

Material	/K	Material	/K
Diamond, Graphite	4000	Silica Glass	1100
Tungsten	3680	Aluminium	933
Silicon Carbide, SiC	3110	Magnesium	923
Molybdenum	2880	Soda Glass	700-900
Niobium	2740	Zinc	692
Alumina, Al_2O_3	2323	Polyimides	580-630[s]
Silicon Nitride, Si_3N_4	2173	Lead	600
		Tin	505
Chromium	2148	Solder (Sn-37Pb)	456
Platinum	2042	Polyesters	450-480[s]
Iron	1890	Polycarbonates	400[s]
Nickel	1726	Polyethylene, high-density	300[s]
Cermets	1700	Polyethylene, low-density	360[s]
Silicon	1683	Epoxy, general purpose	340-380[s]
Copper	1356	Polystyrenes	370-380[s]
Gold	1336	Nylons	340-380[s]
Silver	1234	Polyurethane	365[s]
Acrylic	350[s]	GFRP	340[s]
CFRP	340[s]	Polypropylene	330[s]

Table 2-4. Coefficients of Thermal Expansion at Room Temperature

Material	Coefficient of Thermal Expansion$(\times 10^{-6} K^{-1})$
Silver	19.0
Copper	16.5
Gold	13.8
Aluminium	23.6
Brass	20.0
Tin	23.5
Lead	29.0
Solder (60/40)	24
Nickel	13.3
Alumina	8.8
E Glass	5.5
Silicon	2.5
Aluminium nitride	2.7
Nylon	80-90
Polyethylene	60-220
Polystyrene	50-85
Epoxy resin	26

2.1 The Peculiarities of Tin

Most current solder alloys are tin-based, and it seems likely that the emerging generation of lead-free solders will follow suit. In fact, the popular new alloys contain over 95 per cent tin, whereas the present-day general purpose tin-lead alloy comprises around 60 per cent tin. It is inevitable, therefore, that the properties of the parent metal in the solder alloy will impinge upon those of the solder itself. Amongst engineering metals, tin is unusual in respects which may ultimately affect the performance of a soldered joint in service. First, tin is *allotropic*, which means that its fundamental arrangement of atoms (or unit cells) in the solid state exists in more than one form according to the temperature. Above 13°C, the body centred tetragonal unit cell is known as *white tin*, whereas at lower temperatures it becomes *grey tin* which has a diamond cubic unit cell structure. A substantial volume increase (~27%) and localised cracking are associated with the transition as the temperature falls, and the product is termed *tin pest*. Although the change occurs extremely slowly, prolonged exposure can lead to complete disintegration (Figure 2-1). There are indications that tin pest also forms in dilute tin-0.5 copper alloys used as solders, so in applications involving sub-ambient temperatures for periods of years, this possibility should not be ignored. The presence of significant amounts of lead in conventional solder alloys prevents the transition.

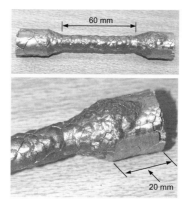

Figure 2-1   Tin pest formed on a sample of Sn-0.5Cu, aged at -19⁰C for 1.5 years.

The second feature of the physical properties of tin which could affect solder performance is its *anisotropy* i.e. within an individual grain, the properties vary according to the direction in which they are measured. Although large numbers of randomly oriented grains, which usually constitute a component or sample, exhibit no directionality, it should be remembered that miniaturisation has resulted in interconnections comprising of few grains. Differences in coefficient of thermal expansion may produce surface roughening and rapid crack initiation during thermal cycling. Figure 2-2 shows the surface of a small sample of pure tin subjected to thermal cycling between 30 and 130°C. Dilute tin-based alloys, such as lead-free solders, behave in a similar manner.

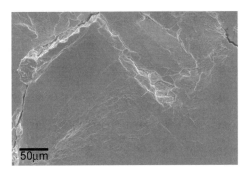

Figure 2-2. Surface damage in a sample of tin, after 3000 thermal cycles between 30 and 130°C

3. MECHANICAL PROPERTIES

In the following section, the fundamentals of mechanical behaviour are introduced, with an emphasis upon the differences between the material classes. The subsequent chapter then considers more complex, but highly relevant, behaviour.

The application of a force, F, to a body of cross section area, A, constitutes a *stress* (designated σ, if the stress acts in tension or compression, or τ, if it is in shear – Figure 2-3). The units of stress are force/area (F/A) and are normally expressed N/mm^2, MN/m^2 or MPa. These are numerically identical. A fixed body will deform under the action of the stress and will be *strained*. For most engineering purposes, strain is determined by the extent of the deformation divided by the original dimension. It is a number only. There are no units. Figure 2-4 illustrates tensile and shear strains, ε and γ respectively.

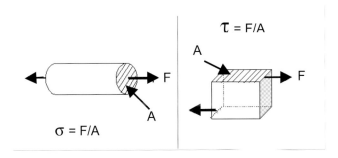

Figure 2-3. Definition of tensile and shear stress

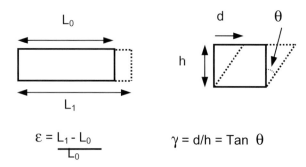

Figure 2-4. Definition of tensile and shear strains

The relationship between the applied stress and the strain it produces is known as a *constitutive equation*. Perhaps, the simplest and best-known response is *elasticity*, when the strain is proportional to stress, and the constant of proportionality is the *modulus*. Modulus has the same units as stress. The constitutive equation for the strain, ε, in the case of tensile/compression behaviour is:

$$\varepsilon = \sigma/E \qquad (2.1)$$

whereas in shear, the strain, γ, is given by

$$\gamma = \tau/G \qquad (2.2)$$

where E and G are the tensile (Young's) and shear modulus respectively. The term *stiffness* is often used for modulus. The higher the modulus or stiffness, the lower is the deformation produced for the same applied stress. Table 2.5 shows the values of modulus for the various materials classes. There is some overlap between ceramics and metals but polymers are considerably less stiff. The values cited in Table 2.5 are typical 'text book' figures. It will be demonstrated later that significantly wide variations exist for solder alloys in the published literature.

Apart from the proportionality between stress and strain, the other principal characteristic of elasticity is that it is *instantaneous* or *time independent*. The applied stress produces an immediate strain that disappears instantly when the stress is removed. (Figure 2-5).

Table 2-5. Elastic Modulus of Materials at Room Temperature

Material	$E/GN\ m^{-2}$	Material	$E/GN\ m^{-2}$
Diamond	1000	Silicon	107
Tungsten Carbide, WC	450-650	Silica Glass, SiO_2 (Quartz)	94
Silicon Carbide, SiC	450	Zinc and Alloys	43-96
Boron	441	Gold	82
Tungsten	406	Aluminium	69
Alumina, Al_2O_3	390	Silver	76
Chromium	289	Soda Glass	69
Magnesia, MgO	250	Tin and Alloys	14
Nickel	214	Fibreglass (glass-fibre/epoxy)	35-45
CFRP	70-200	GFRP	7-45
Iron	196	Graphite	27
Ferritic Steels, low-alloy steels	200-207	Lead and Alloys	14
Stainless austenitic steels	190-200	Polyimides	3-5
Mild steel	196	Polyesters	1-5
Platinum	172	Acylics	1.6-3.4
Boron/epoxy composites	125	Nylon	2-4
Copper	124	PMMA	3.4
Copper alloys	120-150	Polystyrene	3-3.4
Brasses and bronzes	103-124	Polycarbonates	2.6
Epoxies	3	Polypropylene	0.9
Polyethylene, high-density	0.7	Polyethylene, low-density	0.2

All materials exhibit elastic behaviour to varying extents. A key factor in determining the type of strain response to an applied stress is *temperature* – not simply the absolute temperature, but the ratio of temperature to the melting point (or decomposition point in some materials). When these two temperatures are expressed in degrees Kelvin (K), the ratio is known as the *homologous temperature* (T_h). At values of T_h above about 0.4, *time dependent*, recoverable, strain may occur in some polymers and metals. This type of deformation is known as *viscoelasticity*, and is quite common in solders. The homologous temperature at room temperature (20°C) for the common eutectic Sn–37 Pb solder, which melts at 183°C, is 0.64. For the new generation of lead-free solder alloys, which have melting points around 220°C, this ratio falls to 0.60, which is still sufficient for viscoelasticity to occur. Figure 2-5 compares elastic and viscoelastic behaviour. Both strains are completely recoverable, but in elasticity this occurs instantaneously upon removal of the stress.

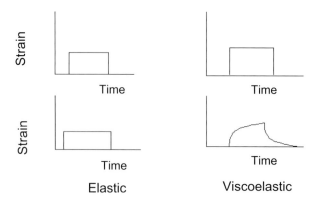

Figure 2-5. Comparison between (a) elastic and (b) viscoelastic behaviour

As stress levels increase, substantial departures in the characteristic responses of the material groups occur. In ceramics, elastic behaviour is followed immediately by cracking and fracture occurs at extremely low strains. This is known as *brittle fracture*. (Figure 2-6a). In many metals, permanent deformation, or *plasticity*, occurs, usually to several tens of percent. This is termed *ductile behaviour*. The stress level at which departure from elastic behaviour becomes apparent is known as the *yield stress*, σ_y, but if this transition point is unclear, then the stress to produce a small amount of permanent strain, typically 0.1 or 0.2 percent is determined, as the *proof stress* (Figure 2-6b). In terms of the applied force divided by the original area, the *engineering stress* reaches a maximum (known as the *tensile strength*, σ_u) and then falls until ductile fracture occurs. The slope of the line between yield and maximum tensile strength varies according to the degree of *work hardening* – which is a measure of the strengthening effect of the prior deformation. The strain to fracture is called the *ductility*, and is a measure of the ability of a material to accommodate deformation – a very relevant property in the performance of solder joints.

Polymeric materials may behave like ceramics or metals according to the temperature and polymer type. Below a critical temperature known as the *glass transition temperature*, T_g, they mirror ceramics in fracturing immediately after elasticity. Above this temperature, they can exhibit substantial ductility although the mechanisms involved are quite different from those producing plasticity in metals. An equivalent *ductile-brittle transition* occurs in many metals but this is usually outside normal service conditions. Figure 2-6 summarises the various types of behaviour that could occur in materials associated with soldered interconnections. It is unusual in engineering to find such a broad spectrum of possibilities in a single application.

Materials that fail in a non-ductile manner are very susceptible to the presence of defects or cracks. Because there is no plastic deformation, stresses concentrate at the tips of pre-existing surface or interior defects. Such flaws serve as crack nuclei, and when stress levels attain a critical value, crack propagation results. Failure is therefore caused by propagation of a crack as distinct from yielding. A measure of a material's ability to withstand stress when it contains a defect is known as its *toughness*. A property called the *plane strain fracture toughness*, $K_{IC,}$ denotes this value as the critical *stress intensity*, K, at fracture (Figure 2-7).

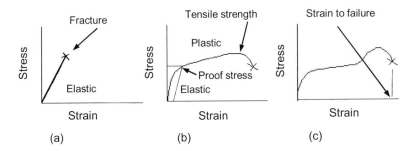

Figure 2-6   Overview of stress-strain responses (a) brittle behaviour of a ceramic, a polymer below its glass transition temperature, or a non-ductile metal, (b) a ductile metal, (c) a polymer above the glass transition.

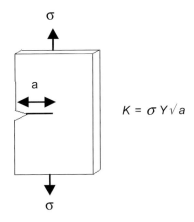

Figure 2-7. Stress intensity at a crack tip

The suffix I relates to the opening mode of the crack. Mode I is tensile opening, whereas Modes II and III describe in-plane and out-of-plane shear opening respectively. Stress intensity is defined as:

$$K = \sigma Y \sqrt{a} \qquad (2.3)$$

where Y is a geometrical constant, σ the applied stress and a is the crack length. The units are MPa$\sqrt{m}$ (or MNm$^{-3/2}$) and very low values exist for ceramics and glassy polymers. (Table 2.6).   So, the strength of these materials falls as the defect size increases.   In electronics equipment, bending stresses or impact damage may cause fracture, either if the stresses are too high or the defects are too large.

Table 2-6. Fracture Toughness Values of Materials

Material	Fracture Toughness $MNm^{-3/2}$
Ductile metals (Cu, Ag, Al, solders)	80-350
Aluminium Alloys	20-45
Fibreglass	42-60
Polypropylene	3
Polyethylene	1-2
Nylon	3
Epoxy	0.3-0.5
Silicon Carbide	3
Alumina	3-5

When a material is strained to a point above yield, and the strain is held constant, the applied stress necessary to maintain the strain falls as a function of time. This is known as *stress relaxation* (Figure 2-8), and it arises as *anelastic recovery* (time-dependent elastic deformation) and, subsequently, internal *creep* (time-dependent permanent deformation) processes occur. It may have important consequences on performance during periods of constant temperature, since damage may be accumulating although the overall strain is essentially constant.

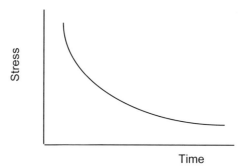

Figure 2-8. Stress relaxation following plastic strain

3.1 Effects of Temperature and Strain Rate

As temperature increases, there is a general trend in metals and polymers, above their glass transition temperature, for their resistance to deformation and their ability to carry load to diminish. Values of modulus and strength decrease, whereas ductility (strain to failure, ε_f) may increase slightly or remain unaffected. (Figure 2-9). The mechanical properties of ceramics are usually insensitive to temperature until well above the domain at which electronics equipment operate - so for the present purposes they can be considered constant.

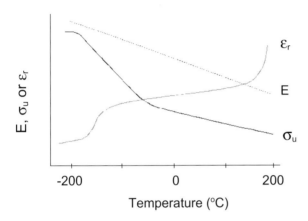

Figure 2-9. Schematic representation of the effect of temperature on mechanical properties of solders

Even when the temperature is fixed, measured values of strength and ductility may change significantly according to the rate at which the force is applied (generally the *strain rate*, $\dot{\varepsilon}$, is considered). Strain rate is the change of strain as a function of time, and is presented in units per second or per hour. For example, a strain rate of 10% per hour ($0.1h^{-1}$) is equivalent to a rod of length 10mm extending to a length 11mm over a period of one hour. At high strain rates, the measured strength is generally larger than that determined at slow strain rates, and, quite often, the ductility is reduced. (Figure 2-10).

Both these facts present an enormous challenge to accurate computational modelling of solder joint behaviour. Since failure in practice generally involves temperature cycling, frequently between limits in excess of 100°C, then it is essential to know the variation of mechanical properties with temperatures in this range. Secondly, it is also important to be aware of the rate of temperature change, which determines the strain rate experienced by the joint, and select the appropriate values of mechanical properties accordingly. The absence of reliable and appropriate data has been a major impediment to the full exploitation of the computer power that is now available for modelling and analysis.

3.2 Mechanisms of Deformation

This section presents a brief review of the internal processes which result in the deformation of materials discussed above. For more detail, any standard textbook on materials science is recommended.

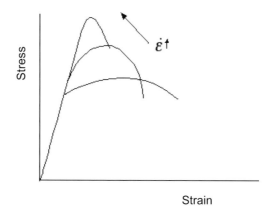

Figure 2-10. Effect of strain rate on stress-strain response (schematic)

Most engineering materials in the solid state are crystalline, in the sense that their atoms are arranged in regular locations. There are repeating configurations known as *unit cells*. For example, many common metals, such as silver, copper, aluminium and iron have a cubic structure. At room temperature, tin has a *body centred tetragonal* structure. Polymeric materials are usually non crystalline although their long molecular chains may sometimes possess a regular arrangement. This occurs on a substantially larger size scale than does the regularity in atomic arrangement in metals and ceramics in which the unit cells occupy a few nanometres.

In simple terms, elasticity involves small displacements of atoms from their original positions. On removal of the applied force, they return instantaneously to these locations. In contrast, plastic behaviour is associated with the sliding of planes of atoms across each other, involving *dislocations*. This produces permanent deformation even when the applied force is removed. The only means to return to the original position involves applying the force in the opposite direction. Viscoelastic and viscoplastic deformation usually occurs in polymers above their glass transition temperature and in metals at temperatures in excess of $0.4\ T_h$. These mechanisms are time-dependent, involving straightening or sliding of long chain molecules or diffusion in metals.

Even crystalline materials are not totally perfect and contain defects. The most significant of these in the present context is the *vacancy,* which is simply an empty lattice site – or a hole where an atom would be expected to be. Vacancies are equilibrium defects and exist at all temperatures except absolute zero. They have a major role in diffusion, which involves an interchange between the atom and the vacancy. Since the equilibrium vacancy concentration increases exponentially with temperature, so does the rate of diffusion. The vacancy facilitates *dislocation climb* around obstacles to their movement on slip planes, and hence plays a major role in work hardening, recovery and creep.

3.3 Modelling Material Behaviour

Once behaviour ceases to be elastic, the relationship between stress and strain becomes more complex. Constitutive equations must describe the various stages of the deformation process. For example, in the case of a metal being deformed plastically, the linear elastic region is generally followed by a 'curved' plastic line. Traditional analysis assumes that either the subsequent strain after yield occurs with no increase in stress - *ideal elastic-plastic behaviour,* or that the curve between the yield stress and the tensile strength may be approximated by a straight line, *linear elastic - plastic behaviour* (Figure 2-11). The slope of this line is the *work hardening coefficient,* m. In this case, the stress-strain relationships may be written as

$$\sigma = \sigma_y + m\left(\varepsilon - \frac{\sigma_y}{E}\right) \tag{2.4}$$

or

$$\varepsilon = \frac{\sigma_y}{E} + \left(\frac{\sigma - \sigma_y}{m}\right) \tag{2.5}$$

More complex mathematical relationships can be employed to describe the curved nature of the plastic response, eg the Ramberg-Osgood expression

$$\varepsilon = \frac{\sigma_y}{E} + \left(\frac{\sigma - \sigma_y}{a}\right)^b \tag{2.6}$$

where b, the strain hardening exponent, and a are constants.

Appreciation that the values of σ_y, σ_u, a, b and m are all sensitive to temperature and strain rate provides an insight of the true complexity of, and the requirements for, reliable modelling. Relative to the service situation, the equations above describe simple conditions. Constitutive equations for more realistic, but complex, behaviour are presented in the following chapter.

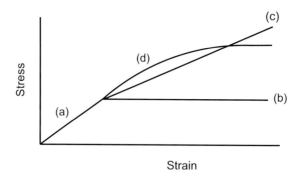

Figure 2-11. Constitutive stress-strain relationships (a) elastic; (b) perfectly plastic; (c) linear work hardening; (d) realistic work hardening

4. CONCLUSIONS

Interconnection regions in electronics equipment may comprise examples of all material classes. The widely different properties, that such materials exhibit, may have a profound influence upon performance.

It is important to appreciate that there is no such thing as a fixed value of a mechanical property of a material. Values may change significantly with temperature, strain rate and, from the material perspective, with microstructure. This variability has gone largely unrecognised in many modelling activities, and the incorporation of realistic property values, which take account of these factors, represents a major challenge to reliable design of solder joints.

5. RECOMMENDED GENERAL READING

Engineering Materials: An Introduction to their Properties and Applications, M F Ashley and D R H Jones, (Pergamon Press), 1980.
Deformation and Fracture Mechanics of Engineering Materials, R W Hertzberg, (John Wiley & Sons), Fourth Edition, 1996.
Mechanical Metallurgy, G E Dieter, (McGraw-Hill), 1988.

CHAPTER 3

MORE COMPLEX MECHANICAL BEHAVIOUR

1. INTRODUCTION

In this chapter, the range of possibilities relating to how a force (or stress) can be applied is extended. A body can be stressed very rapidly as in *impact*, or over a period of months or years, leading to *creep* failure. Alternatively, it may be subjected to repeated stresses (or strains) until it fails by *fatigue*, a process that may be mechanically or thermally induced (*thermal fatigue*). All of these processes may occur in printed circuit boards, and affect reliability. The key parameters in these more complex, but often more pertinent, modes of failure are now considered, and the common methods of data representation for the designer are reviewed.

2. IMPACT

Electronic equipment is never intentionally subjected to rapidly increasing forces, although it may experience such conditions by accident when, for example, the equipment is dropped. If this happens, it is those materials that are already vulnerable to brittle fracture, such as ceramics and polymers below their glass transition temperatures, which are most likely to fracture. Toughness values and the quality of the material, in terms of defect size and distribution, are the key parameters here. Electronic assemblies are usually designed to minimise the transmission of impact forces to the vulnerable components. Ductile materials can usually accommodate instantaneous loading, although the permanent deformation induced in them may cause subsequent problems. So, overall, impact failure should be uncommon, and interest in the values of impact resistance is not generally a high priority.

3. CREEP

The converse of impact is the situation in which a body is loaded to a fixed value and left in that condition for a prolonged period. In some applications, this may involve many years. If the limit of elastic behaviour is exceeded, the material will continue to deform under the action of the constant load (or stress), and this process is known as creep.

The designer encounters most problems from the creep of metals, or polymers above their glass transition temperature. Again, temperature is a critical parameter, and for metals, creep must be considered when values of the homologous temperature exceed about 0.4. For the conventional Sn-37Pb solder, this is equivalent to $-90°C$, which means that the creep behaviour of these solders should be considered in all applications. The generally higher melting points of the new lead-free solder alloys raises this critical temperature to $-75°C$ which does not materially alter the situation.

Creep is usually represented as a graph of strain, ε, against time, t, and it continues until the specimen ruptures at a time, t_r, (Figure 3-1). In metals, there are three stages of creep, although all stages are not necessarily apparent in any single creep curve. After initial

33

instantaneous elastic deformation, ε_0, on loading, *primary creep* is denoted by a continuous fall in creep strain rate. *Secondary creep* is characterised by a period in which the creep strain rate remains essentially constant, and in *tertiary creep* the creep strain rate increases continuously until rupture. Secondary creep is sometimes described as *steady state* creep, and the value of the creep strain rate during this stage is the minimum for the test and that used in many design calculations.

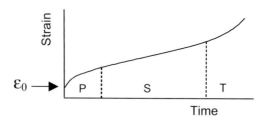

Figure 3-1. Schematic strain v time graph for creep, indicating the stages of creep (P = Primary; S - Secondary; T = Tertiary)

The graph of creep strain *v* time is not very convenient for displaying the results of numerous creep tests. This is best achieved by plotting applied stress against time to rupture (Figure 3-2), when the focus of attention is rupture time rather than the amount of creep strain or the rate at which it accumulates. It is easier to use logarithmic scales for each axis so that an enormous range of information, up to 10^5h in some cases, can be presented in a single figure. Such graphs often appear linear or bi-linear, and the existence of transition is indicative of a change in the dominant creep mechanism. These features present a great challenge to reliable life prediction, since it is not unusual for laboratory-produced data to be in one dominant creep regime whereas the actual service condition (usually lower stress, longer term) is governed by different creep processes.

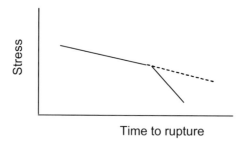

Figure 3-2. Stress v time to rupture for creep illustrating a change in dominant creep mechanism

Linearity in a plot of log stress *v* log time to rupture may often be masked by significant changes in creep ductility over the stress range examined. An unequivocal means of determining the existence of a mechanism change is by plotting secondary or minimum creep strain rate, $\dot{\varepsilon}_m$, against applied stress σ, on logarithmic scales (Figure 3-3). Commonly, the relationship

$$\dot{\varepsilon}_m = B\sigma^n \qquad\qquad (3.1)$$

is indicated. (Norton's Law). A change in gradient denotes the transition and its location. For solders, it is not uncommon for more than one transition to be apparent on this graph, which renders modelling and life prediction even more challenging.

Figure 3-3. Minimum (or steady state) creep rate v applied stress indicating changes in dominant mechanism

So far, only creep rupture has been considered, but in some circumstances a component may 'fail' after a certain amount of deformation at times well below that required for rupture to occur. In this case, rupture lives may be replaced by 'time to achieve a specific amount of deformation'. Conversely, while it is tempting for the designer to select materials on the basis of the greatest time rupture, creep ductility must also be considered. Materials that have very low creep ductilities (say below 5 percent) tend to fail suddenly without warning (Figure 3-4). As with tensile and shear strength, a compromise is necessary. Deformation-based failure criteria require knowledge of the initial and primary creep strains which can be significant.

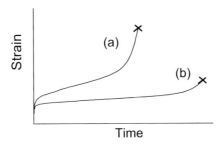

Figure 3-4. Strain v time curves for a ductile material (a) and a creep-resistant, low ductility, material (b)

3.1 Creep Mechanisms

Creep may involve a number of mechanisms of deformation, such as diffusion-controlled dislocation motion or the migration of vacancies, either through the matrix or along grain boundaries.

At stress levels in excess of yield, dislocation motion produces plastic deformation the extent of which is determined by obstacles on the slip plane. Subsequent time-dependent deformation can occur if the dislocation is able to *climb* over the obstacle by acquiring vacancies. Once climb has occurred, the dislocation will be free to move (i.e. no longer pinned by precipitates, solute atoms or other dislocations). Clearly, the rate of climb depends upon the number of vacancies in the lattice, which is proportional to the temperature. The vacancy concentration, v, is $\infty(-E_f/kT)$ where E_f is the energy of vacancy formation and k is the Boltzmann constant. Therefore, the rate of dislocation climb and creep increases exponentially with temperature.

At stresses lower than those required for dislocation glide, creep can occur via diffusion of vacancies alone. The *Nabarro-Herring* creep mechanism dominates the creep process at low stress levels and high temperatures. This mechanism is accomplished solely by diffusional mass transport in the bulk of the material. It occurs at temperatures greater than $0.7T_m$, when there are more diffusional paths in the bulk than in the grain boundary. A higher vacancy concentration exists in the regions of a material experiencing a tensile stress in comparison to those subject to a compressive stress. This produces a vacancy flux from the regions of material under tension to those regions under a compressive stress. Therefore, there will be a mass flux in the opposite direction. This results in an elongation of the individual grains along the tensile axis and is equivalent to a creep strain (Figure 3-5).

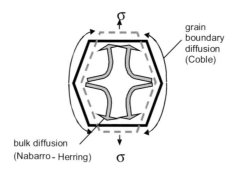

Figure 3-5. Mechanisms of diffusion creep

Coble creep is driven by the same vacancy concentration gradient as Nabarro-Herring creep, although mass transport occurs by diffusion along the grain boundaries in a polycrystalline material or along the surface of a single crystal. This is caused by a lower activation energy requirement than for bulk diffusion. Coble creep is dominant at temperatures lower than those for Nabarro-Herring creep. In this case, the temperature is too low for bulk diffusion, and diffusion is activated in the grain boundary and, consequently, Coble creep will dominate in very fine-grained materials when more diffusion paths are available. As these two mechanisms are parallel creep processes, the overall creep rate, $\dot{\varepsilon}_t$, due to diffusional flow is the sum of each process, as represented below:

$$\dot{\varepsilon}_t = \dot{\varepsilon}_{NH} + \dot{\varepsilon}_C \qquad (3.2)$$

where $\dot{\varepsilon}_{NH}$ and $\dot{\varepsilon}_C$ are the contributions from Nabarro-Herring and Coble creep respectively.

Table 3.1 summarises the various domains of creep and typical values of n and Q associated with them.

Table 3-1. Variation of creep exponent and activation energy with stress and temperature. Where Q_{SD}, Q_{core} and Q_{GB} are the activation energies for self diffusion, core diffusion and grain boundary diffusion respectively.

Creep Process	Temperature	Stress	n	Q
High temperature dislocation creep	Above 0.7 T_m	Intermediate/High	>3	~Q_{SD}
Low temperature dislocation creep	~ 0.4 – 0.7 T_m	Intermediate/High	>3	$Q_{core\ (pipe)}$
High temperature diffusional creep	Above 0.7 T_m	Low	~ 1	Q_{SD}
Low temperature diffusional creep	~ 0.4 – 0.7 T_m	Low	~ 1	Q_{GB}

3.2 Creep Life Prediction

As indicated previously, simple extrapolation of laboratory data out to service lifetimes is hazardous, especially in view of potential changes in dominant creep mechanism. Numerous approaches exist for creep life prediction - some quite complex and none regarded as universally applicable.

A popular and simple approach is based upon the Monkman-Grant equation,

$$\dot{\varepsilon}_m t_r = C \qquad (3.3)$$

where $\dot{\varepsilon}_m$ is the minimum creep strain rate, t_r is the time to rupture and C a constant related to the creep ductility. The minimum creep rate may be linked with the applied stress, σ, by various equations according to the dominant creep mechanism, e.g.

$$\dot{\varepsilon}_m \propto \sigma^n \quad \text{(power law creep)} \qquad (3.4)$$

$$\dot{\varepsilon} \propto \exp(a\sigma) \text{ (exponential creep)} \tag{3.5}$$

$$\dot{\varepsilon}_m \propto \left[\sinh(b\sigma)\right] \text{ (combination creep)} \tag{3.6}$$

where a and b are temperature dependent constants. By adjusting the constants, the same form of these equations can be written for the rupture time, t_r.

The Monkman-Grant expression is based upon the minimum, or steady state, creep rate, and its accuracy is influenced by the proportion of the entire creep life that this stage occupies. This is best revealed by plotting current strain rate, $\dot{\varepsilon}$, against fraction of life (Figure 3-6). The proportion of life occupied by steady state creep may vary between 100 and 0 per cent (Lines (a) and (c) respectively). An intermediate situation illustrated by (b) is more common. For most solders, the fraction of life for which the strain rate is within 10 per cent of its minimum value, is around 25-30 per cent (1). Given the potential for the most influential mechanism to change, and the vulnerability of the Monkman-Grant equation, then the uncertainty surrounding creep life prediction can be appreciated. This challenge is examined in more detail in Chapter 12.

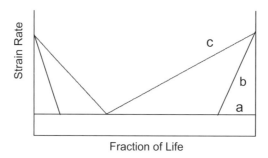

Figure 3-6. Plot of incremental strain rate as a fraction of creep life, indicating the variable significance of the secondary (or steady state) stage

3.3 Effect of Temperature and Stress Level

Creep is highly sensitive to both applied stress level and to test temperature. Figure 3-7 illustrates the general trend for both to promote creep as they increase. As a thermally activated process, creep rate increases exponentially with temperature. Typically, an increase in temperature of 20K can double the creep rate and halve the creep life. The effect of stress is more complex and is dependent upon the controlling creep mechanism as indicated in Figure 3-3. Values of the stress exponent, n, may vary between 1 and over 15, which again re-inforces the need to identify the dominant mechanisms in service. A popular generalised expression for the steady state creep rate, $\dot{\varepsilon}_m$, is

$$\dot{\varepsilon}_m = A\sigma^n g^{-p} \exp\left(\frac{-Q}{RT}\right) \qquad (3.7)$$

where g is the grain size, A, n and p are constants, and Q is the activation energy of the dominant creep process. Determination of the value of Q enables identification of the controlling creep mechanism to be made. This is achieved by plotting log minimum creep rate at a constant stress level versus reciprocal of temperature (K), when the gradient of the line is −Q/R.

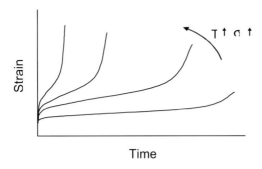

Figure 3-7. Effect of stress and temperature on the creep strain-time graph

4. FATIGUE

Failure due to the repeated application of a number of stresses (or strains), the magnitude of which is insufficient to cause failure when applied singly, is known as *fatigue*. The process involves the initiation and gradual growth of cracks until the remaining section of the material can no longer support the applied load. It is by far the most common mode of failure in all engineering applications.

The designer is interested in the number of cycles of stress which cause failure, and this is principally determined by the stress range, $\Delta\sigma$, ($\Delta\sigma = \sigma_{max} - \sigma_{min}$) - or the equivalent strain range, $\Delta\varepsilon$, ($\varepsilon_{max} - \varepsilon_{min}$) if the body is being cycled between limits of strain, rather than stress. Under conditions of stress control, the traditional means of conveniently presenting data and comparing performance is via an S – N graph, i.e. a plot of stress range versus number of cycles to failure (Figure 3-8). As might be expected, the smaller the stress range, the larger the number of cycles required for failure. A few materials, notably steels, exhibit a *fatigue limit*, which is a value of the stress range below which fatigue failure does not occur. The majority of materials display an increasing *fatigue endurance* as the stress range diminishes. Data in this form are difficult to analyse or to use in expressions for life prediction. A property more fundamental to materials behaviour than stress is strain, because that is a measure of the extent to which the material is deformed. Plots of strain range against numbers of cycles to failure are considerably more useful to the designer, particularly when plotted on logarithmic scales, when linear relationships often exist, as indicated in the following paragraph.

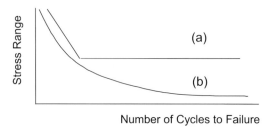

Figure 3-8. Traditional S-N (stress range v number of cycles to failure) plots for materials, (a) with and (b) without a fatigue limit

It has been demonstrated previously that elastic strains extend usually to a fraction of one percent in metallic materials. After that, permanent deformation (plasticity in metals) occurs. The total strain in this case, ε_t, is made up of an elastic component, ε_e, and a plastic component, ε_p. Similarly, in terms of strain ranges,

$$\Delta\varepsilon_t = \Delta\varepsilon_e + \Delta\varepsilon_p \tag{3.8}$$

A plot of log $\Delta\varepsilon_t$ v log N_f is a curve as shown in Figure 3-9. However, under conditions where either $\Delta\varepsilon_e$ or $\Delta\varepsilon_p$ dominate, plotting these terms against log N_f gives a linear relationship. Hence, either

$$\Delta\varepsilon_e N_f^{\alpha} = C_1 \text{ (Basquin), or} \tag{3.9}$$

$$\Delta\varepsilon_p N_f^{\beta} = C_2 \text{ (Coffin-Manson)} \tag{3.10}$$

where α, β, C_1 and C_2 are constants.

For solders, the extent of elastic strain is small, so that in most cases, $\Delta\varepsilon_p \gg \Delta\varepsilon_e$, and a graph of either $\Delta\varepsilon_t$ or $\Delta\varepsilon_p$ against number of cycles to failure is linear. This is the *Coffin-Manson* expression that is found in some form in many life predictive methods for solder joints.

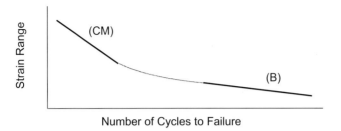

Figure 3-9. Strain range v number of cycles to failure showing the domains of elastic (Basquin) and plastic (Coffin-Manson) dominance

4.1 Hysteresis Loops

A convenient means of depicting stress-strain relationships during fatigue is via a *hysteresis loop* that considers all four quadrants of stress and strain. Figure 3-10 shows initial loading from the origin in the positive – positive quadrant, followed by plastic deformation up to a predetermined strain limit, ε_{max}. The direction of straining is reversed until it reaches a value of $-\varepsilon_{min}$. Note that on reversal, elastic behaviour occurs followed by yield and plasticity in the reverse direction. The 'width' of the loop is equal to the plastic strain range, $\Delta\varepsilon_p$, and the elastic strain range, $\Delta\varepsilon_e$ is given by the difference between this and the total strain range. The area enclosed by the loop is a measure of the energy required to produce the observed deformation. This type of representation is very convenient when considering more complex and realistic cycles.

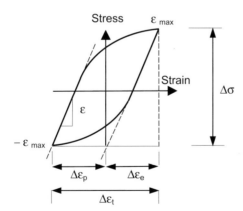

Figure 3-10. Mechanical hysteresis loop showing the complete stress-strain relationship during a fatigue cycle

When cycled between strain limits, materials may respond in a variety of ways before crack initiation and growth. Some get stronger, or *cyclically harden*, which means that greater applied stresses are required to attain the desired strain limits (Figure 3-11). Pure metals, stainless steels and lead-rich solders fall into this category. In contrast, other materials, such as precipitation hardening alloys, the eutectic lead-tin and many lead-free solders *cyclically soften,* in that they lose strength when subjected to strain cycling. A third group, typified by nickel and titanium alloys, is *cyclically stable.* The final fall in stress range in all cases is associated with the formation of a macroscopic crack that impairs the load bearing capacity of the specimen. It should be noted, however, that cyclic response in a single material can change according to the temperature and the applied strain range.

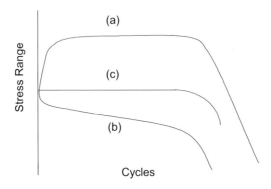

Figure 3-11. Variation of stress range with cycles necessary to maintain constant strain limits. (a) cyclic hardening, (b) cyclic softening, (c) cyclic stability

4.2 Fatigue Crack Growth

As the capability of non-destructive evaluation has improved, the existence of defects in materials has been recognised as inevitable. A new methodology for design, *defect tolerance*, has emerged. The central question is how long (or how many cycles of stress or strain) will it take a pre-existing defect to grow to catastrophic dimensions under service conditions. In Figure 3-12, this is equivalent to a crack growing from an initial size, a_o, to a catastrophic dimension, a_t. The value of a_t is determined by the fracture toughness of the material.

Under fatigue, the crack growth rate, da/dN, may be correlated with the crack tip stress intensity range ΔK, by

$$\frac{da}{dN} = C(\Delta K)^m \qquad (3.11)$$

where m and C are constants. Integration of this expression, between the initial crack size and the critical dimension at failure, provides an estimate for the number of cycles to failure, N_f. This approach is based upon linear elastic fracture mechanics (LEFM) which implies limited plasticity, confined to the crack tip. In this form, its application to interconnections is impaired by the ductile nature of solders and the existence of short cracks, the growth of which often does not follow the above expression. However, substituting a plastic equivalent of K, such as the J integral or an energy term based upon the area of the hysteresis loop, has proved successful. In actual joints, there is the added complication of the intermetallic compound that may affect both initiation and growth of cracks. Chapter 13 is devoted to life prediction based upon crack growth measurements.

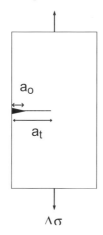

Figure 3-12. Schematic illustration of the defect tolerant approach to life prediction

4.3 Dwell Periods During High Strain Cycling

Components rarely experience stresses or strains that vary in a continuous and regular manner. There are generally periods in which stress or strain levels are fairly constant, such as take-off, flight and landing of an aircraft, or power-on, operate and power-off of an electronic device. The key question in such cases is 'What is the effect of hold or dwell periods under constant conditions on the number of cycles to failure?

During strain-controlled fatigue at temperatures in excess of $0.4\ T_h$, the presence of a hold period is associated with stress relaxation – or fall-off, as internal creep processes convert elastic strain to creep strain, and this is associated with the requirement for less externally applied stress to maintain the constant strain limits. Dwells may exist at any position on the hysteresis loop but are usually applied at the strain extremes as shown in Figure 3-13. A convenient notation for dwell-containing cycles is; o/o – continuous cycling, no dwell; t/o – dwell at maximum (tensile) strain; o/t – dwell at minimum (maximum compressive) strain, t/t – dwells at both extremes. Under certain conditions, the presence of a dwell can be profoundly influential on fatigue endurance, as will be seen later. This situation is often described as a *fatigue-creep interaction*, although a physical interaction between the damage processes is the exception rather than the rule.

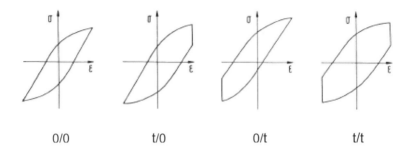

0/0   t/0   0/t   t/t

Figure 3-13. Mechanical hysteresis loops indicating location and effect of dwell periods in the fatigue cycle

5. THERMOMECHANICAL FATIGUE

The origins of the forces in fatigue may be mechanical, such as external loads or fluctuating vibrations, or thermal. In the latter instance, the stresses and strains develop due to differing amounts of expansion or contraction during temperature changes. This damage may build up on a microscale, when different phases within the microstructure have different expansion coefficients, or on a macroscale when materials with different coefficients of thermal expansion are joined to each other, as in the case of the solder joint. The resulting *thermomechanical fatigue* is the principal mechanism of failure in service. Strictly speaking, there are two forms of this process. The application of fluctuating temperatures to a body in which internal constraints to free expansion exist (such as different phases in a microstructure) is known as *thermal stress fatigue (TSF)*. In contrast, a body exposed to temperature fluctuations and for which free expansion is limited by external constraints, experience *thermomechanical fatigue* (TMF). A key difference is that in TSF, analysis of the stress-strain conditions is difficult, whereas in TMF these parameters may be independently controlled, making analysis relatively more straightforward. Very often in the literature, this distinction is not made, and the term thermomechanical fatigue is used universally.

5.1 Development of Strains During Thermal Cycling

It has been shown previously that the coefficients of thermal expansion, α, of the materials involved in solder joints may vary considerably (Table 2.4). A factor of five or more is possible. When two materials are in contact, a change in temperature will produce different displacements in each material. A strain will be generated. This can be modelled by considering two pieces of material as shown in Figure 3-14.

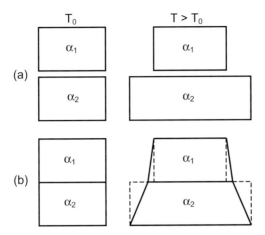

Figure 3-14. Thermal strains developed between dissimilar materials (coefficients of expansion α_1 and α_2 in contact (a) free expansion; (b) materials in contact

Assuming that equilibrium (zero stress, zero strain) exists at temperature, T_0, a temperature change, ΔT, will induce a strain, ε_T, given by:

$$\varepsilon_T = \Delta T(\alpha_1 - \alpha_2)$$ (3.12)

where α_1 and α_2 are the coefficients of thermal expansion of the two materials. Even with large differences in coefficients of thermal expansion (say 20 x $10^{-6}\,K^{-1}$), and a value of ΔT of 100°C, the resulting strain would be fairly small (≈ 0.2 per cent). However, in an actual solder joint there is a 'leverage' factor of the material between adjacent solder joints. Figure 3-15 illustrates this with a model joint comprising bars of two materials joined together by solder separated by a distance, L, and by a vertical distance, h. Here, α_s and α_c are the coefficients of thermal expansion of the two materials respectively; T_0, T_{min} and T_{max} are the equilibrium, minimum and maximum temperatures. If $\alpha_s > \alpha_c$, the lower bar contracts more on cooling, and expands more on heating.

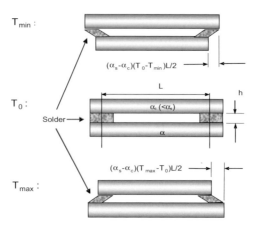

Figure 3-15. Schematic of shear strain due to thermal mismatch in solder joints

The respective relative displacements are:

$$\left(\alpha_s - \alpha_c\right)\left(T_0 - T_{min}\right)\dfrac{L}{2} \text{ on cooling} \qquad (3.13)$$

and

$$\left(\alpha_s - \alpha_c\right)\left(T_{max} - T_0\right)\dfrac{L}{2} \text{ on heating} \qquad (3.14)$$

Since the shear strain, γ, is displacement/height, then for a temperature excursion to T_{min}:

$$\gamma = \dfrac{\left(\alpha_s - \alpha_c\right)\left(T_o - T_{min}\right)L}{2h} \qquad (3.15)$$

Similarly, for a complete thermal cycle between T_{min} and T_{max}, the shear strain range, $\Delta\gamma_t$, is given by:

$$\Delta\gamma = \dfrac{\left(\alpha_s - \alpha_c\right)\left(T_{max} - T_{min}\right)L}{2h} \qquad (3.16)$$

This exceeds the strains developed in the earlier example of two blocks of materials in total contact by a factor of $L/2h$. Depending on the relative magnitudes of L and h, this can result in strains in excess of ten percent. Practically, the joint height is very influential. As indicated earlier, the elastic behaviour of most metallic materials is rarely greater than 0.1 – 0.2 percent. Therefore, temperature cycling can easily produce high strain fatigue conditions, in which the plastic strain range, $\Delta\gamma_p$, is much greater than the elastic strain range component, $\Delta\gamma_e$.

5.2 Hysteresis Loops Produced By Thermomechanical Cycling

During continuous cycling under strain control at a constant temperature, it has been shown that the complete stress-strain relationship may be represented by a hysteresis loop symmetrical about the origin. The width of the loop is a measure of the plastic strain range. The insertion of a dwell period in a cycle results in an asymmetric loop. When both temperature and strain are cycled, the situation is more complex. There is generally an asymmetry about the stress axis due to the variation of the material's strength with temperature.

Strain and temperature may be varied *in-phase* (peak temperature with peak strain) or *out-of-phase* (peak temperature with minimum strain). Idealised loops shown in Figure 3-16 indicate how the TMF loop is essentially the product of the isothermal loops at minimum and maximum temperature (2). A compressive mean stress is generated for in-phase cycling, and a tensile mean stress for out-of-phase fatigue (3). Under conditions where there is significant cyclic plasticity, as is likely in solder alloys, mean stresses rapidly disappear and have little effect on endurance (4). Figure 3-17 shows the hysteresis loops for a tin-silver eutectic solder at 25 and 80°C, and for TMF between those temperatures (5).

Figure 3-18 illustrates schematically how the inclusion of a dwell period at maximum tensile strain in a cycle affects the hysteresis loop (6). During isothermal cycling, the stress increases progressively until the maximum strain limit is reached. On strain reversal, unloading occurs until the minimum strain is attained. With in-phase cycling, the stress level may exhibit a maximum value and then fall because the temperature dependence of strength means that at the higher temperature, a smaller stress is required to maintain the desired strain level. The extent of stress relaxation during the hold period is shown to be similar in each case. Since strength usually increases with diminishing temperature, a similar inflection is not apparent on strain reversal to low temperatures. The existence, location and extent of inflections in the hysteresis loop are governed by the sensitivity of strength to temperature, the work hardening characteristics and their temperature sensitivity, and the strain (or temperature) range of the test. With low strength, ductile, materials such as solder, hysteresis relationships quite distinct from the classic loop may be observed.

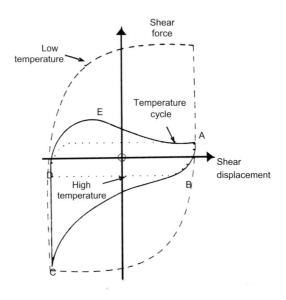

Figure 3-16. The evolution of a TMF hysteresis loop from the isothermal loops at the peak temperatures (2)

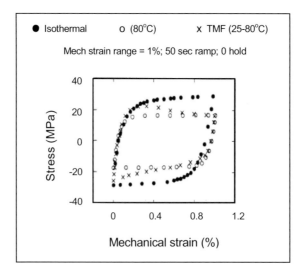

Figure 3-17. Comparison between isothermal and TMF Hysteresis loops in tin-3.5 silver alloy (5)

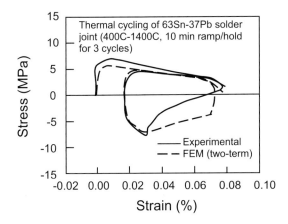

Figure 3-18. Characteristic TMF hysteresis loop for cycle containing hold periods (6)

6. CONCLUSIONS

When actual service conditions are considered, more complex mechanical properties are influential in determining performance. Time-dependent deformation (creep) may take place by several mechanisms, each governed by different parameters and expressions. Fatigue endurance may be affected by the presence of dwells in the strain-time cycle, and the specific failure process, involving crack growth, may be strongly influenced by intermetallic compounds within the joint and the presence of existing defects. These effects may be different in different solder alloys. During thermal cycling, the situation becomes more complex when the mechanical properties change continuously with temperature. Substantial plastic strains can be produced in the solder. Only by obtaining a thorough understanding of the dominant failure mechanisms under specific service conditions, and by utilising mechanical properties and constitutive equations associated with those conditions, can reliable life prediction be achieved.

7. REFERENCES

1   WJ Plumbridge, Soldering and Surface Mount Technology, 2003, 15, 26.
2   PM Hall, IEEE Trans on Components, Hybrids and Manufacturing Technology, 1984, 7, 14.
3   K Kuwabara, A Nitta and T Kitamura, Proc. Int. Conf. on Advances in Life Prediction methods, ASME (New York), 1983, 131.
4   BI Sandor, 'Fundamentals of Cyclic Stress and Strain', (Univ. of Wisconsin Press), 1972, p.106.
5   M Mavoori, S Vaynman, J Chin, B Moran, L Keer and ME Fine, Materials Research Soc. Symp., 1995, 390, 161.
6   YH Pao, S Badgley, R Govila and E Jih, Fatigue of Electronic Materials, ASTM STP 1153, (ASTM Philadelphia), 1994, 60.

CHAPTER 4

MECHANICAL TESTING (FOR ELECTRONICS APPLICATIONS)

1. INTRODUCTION

Having examined the various modes of mechanical behaviour, attention is now turned to the evaluation of these properties. With the increased focus upon structural integrity, reliable information on relevant properties determined under realistic conditions is an essential precursor to efficient design and life prediction.

A soldered interconnection is a quite complex structure (Figure 4-1). On the macroscale, it comprises different materials with significant variations in properties between them, and on the microscale, the microstructure of the solder alloy is not only unstable and highly sensitive to thermal history, it generally contains intermetallic compounds (IMCs) which themselves can have a profound effect on performance. Evaluation of the properties of a joint is highly specific to the specimen being measured. At one extreme, there is the bulk solder property and, at the other, is the actual joint that comprises the complex structure described above.

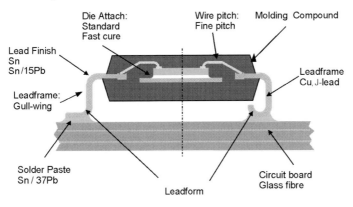

Figure 4-1. Schematic view of a soldered interconnection

In this chapter, the fundamentals of mechanical testing *bulk* samples are first reviewed. The properties obtained may be regarded as less variable than those determined from measurements on joints, and therefore more useful in analysis and modelling procedures. Nevertheless, they remain susceptible to temperature, strain rate and microstructural instability effects. Subsequently, mechanical evaluation of solder *joints* is considered. This is an expanding area, since the adoption of structural integrity approaches, as practised in aerospace and power generation industries, is seen as a way of accommodating the pressures of continued miniaturisation and higher performance demands.

2. MODES OF TESTING

Materials may be stretched, twisted, bent or squeezed. More technically, they can be subject to tensile/compressive, torsional, bending or shear forces which can produce characteristic modes of deformation. In electronics equipment, boards may experience bending or twisting, while solder joints are generally sheared. These modes are illustrated schematically in Figure 4-2 that demonstrates the two basic modes of deformation induced by normal, principal stresses (equivalent to tension or compression) and shear stresses.

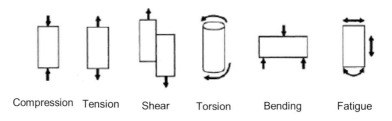

Compression Tension   Shear   Torsion   Bending   Fatigue

Figure 4-2. Various modes of stressing

The correlation between properties measured in tension and shear involves *Poissons ratio*,ν, (which is the ratio of lateral contraction to the extension). For example, the shear strength, τ_u, is related to the tensile strength, σ_u, by

$$\tau_u = \frac{\sigma_u}{2(1+\nu)} \tag{4.1}$$

and the shear modulus, G, is related to the tensile modulus, E, by

$$G = \frac{E}{2(1+\nu)} \tag{4.2}$$

Values for ν are usually in the range 0.25 – 0.35, with 0.33 as a commonly selected figure for calculations involving elastic behaviour and 0.5 for situations in which plasticity dominates. Unlike the samples above, solder joints have irregular shapes and may be subjected to complex stresses. The determination of the location and direction of critical levels requires *stress analysis*.

Monotonic testing involves the application of a gradually increasing force in any mode until failure occurs. Creep may be regarded as a special case, since a fixed force is applied and held constant, and the resulting time-dependent strain is monitored. The converse of creep is stress relaxation, which involves applying a strain to a set limit and measuring the reduction of stress with time. In contrast, cyclic loading is associated with repetitious loading – unloading (or reverse loading) that, again, may be in any of the modes described earlier. Most testing is carried out at a constant temperature, and is described as *isothermal*. Some applications, including electronics, experience fluctuating temperatures, and simulation often involves this.

3. TEST MACHINES

The most elementary means of applying a force to a specimen is by *dead weight loading,* either directly or via a lever mechanism. Load increases are by step increments, so this type of machine is generally restricted to creep testing when a fixed load is applied and remains constant.

For the application of a gradually increasing force, *screw driven* machines have been traditionally employed for monotonic and cyclic testing. The crosshead of the machine is displaced by rotation of the drive columns. With such equipment, displacement rates are limited, especially during fatigue testing when the transition in the direction of displacement is associated with delays and 'flexibility' of the loading system. There are also substantial restrictions on the shape of the displacement-time cycle.

Such shortcomings are largely eliminated in test machines incorporating *closed-loop servohydraulic control systems*, and these are now regarded as standard for toughness, high strain fatigue and crack propagation studies. When service conditions involve prolonged dwell periods (several hours) in strain – time cycles, *electrohydraulic machines* are preferred because of their greater stability. For high cycle fatigue (more than 10^4 cycles) or crack growth studies at low stress ranges, *resonance machines* have been extensively used. However, their operating range and specimen geometry are limited due to the requirement for resonance to occur.

3.1 Control Modes

A specimen in a test machine may be evaluated in one of three common control modes. Selection of the one most appropriate to the conditions experienced in service is desirable.

Position control involves adjusting the position of the crosshead or loading actuator between predetermined limits, and by calibration or assumption, calculating the strains within the gauge length of the specimen. A load cell within the loading train provides stress values by consideration of the cross sectional area of the specimen.

Load control utilises the load cell to apply loads between designated limits, which can be converted to stress.

In *strain control*, a specific gauge length is selected as the operating range of an extensometer attached to the specimen. Displacements of the extensometer provide the required strain levels to enable direct strain control to be achieved. Figure 4-3 summarises these control modes.

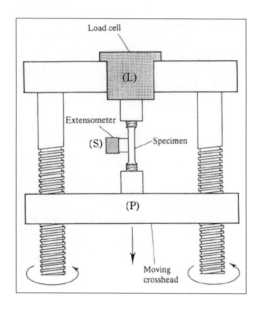

Figure 4-3. Control modes for test machines. (P) crosshead position; (L) load control; (S) direct strain control

The performance of an interconnection in service is usually governed by thermomechanical fatigue when the entire joint experiences thermal fluctuations. Of the materials involved, the solder is often the softest, which results in the strains developed being concentrated there. In effect, the solder is constrained by the adjacent materials, and the situation is often described as being under strain control. Therefore, this mode of control is preferred for fatigue testing of solder alloys, especially when dwell periods are included in the strain-time cycle.

When both temperature and strain are controlled and cycled, the situation is more complicated. Changes in specimen dimensions, thermally induced strains and variations in mechanical properties as a function of temperature must be considered. In order to test to strain limits, determined by the temperature extremes, computer controlled, closed-loop systems are essential.

4. WHAT IS 'FAILURE'?

While a malfunction in electronic equipment is generally associated with a loss of electrical continuity, from the evaluation standpoint a mechanical criterion for failure is usually preferred. Complete fracture or separation is absolute, but it causes problems in testing and has no element of conservatism built in. In cyclic tests, a more popular approach is to examine the stress changes required to maintain the constant strain range limits of the test. The phenomena of cyclic hardening, softening and stability have been introduced previously. When cyclic hardening occurs (as in lead-rich, tin-lead solders) the onset of load fall-off after the saturation plateau is a convenient identification for failure. The situation for alloys which exhibit cyclic softening (such as eutectic Sn-37Pb or many of the lead-free compositions) is more complex. According to the applied strain range, a saturation plateau may, or may not, be apparent in the stress range versus cycles plot. In general, a low strain range is associated with a clearly delineated plateau, but with increasing strain ranges, substantial crack initiation occurs earlier in life and eliminates the stable load range phase. It is usual to adopt a percentage load (or stress range) fall-off from the initial cycle to denote failure. (Figure 4-4). Values such as 10, 20 or 50 percent are common. Which is selected may have a noticeable effect on the Coffin-Manson, or equivalent, graph (Figure 4-5), and the relationship between the failure definition and fatigue damage is unclear.

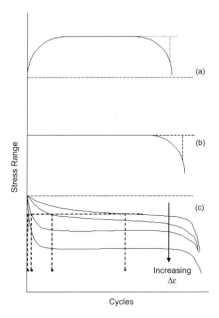

Figure 4-4. Influence of cyclic response on determination of number of cycles to failure.
(a, cyclic hardening; b, cyclic stability; c, cyclic softening)

An alternative method is to identify the point at which crack growth begins to accelerate. This is achieved by plotting the stress range drop per cycle, $d\Delta\sigma/dN$, versus number of cycles, when a clear transition in slopes results. A similar transition may be obtained by plotting the unloading modulus, $d\sigma/d\varepsilon$, immediately after a strain reversal, as a function of cycles.

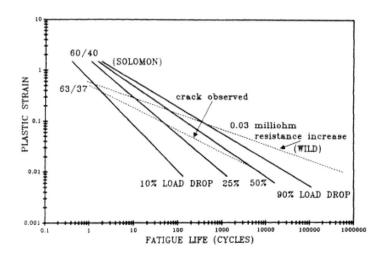

Figure 4-5. Effect of failure definition on Coffin-Manson plot (1)

5. STANDARDS FOR MECHANICAL TESTING SOLDERS

Many international standards exist which prescribe conditions under which mechanical properties, in general, should be determined. This facilitates comparison of results from different laboratories. However, it has already been demonstrated that solders operate at an extreme end of the normal range for engineering materials. The high homologous temperatures encountered in service are responsible for low strength, high strain rate sensitivity and significant microstructural instability. In effect, this means that data not obtained under identical test and material conditions are likely to exhibit a wider scatter than most other engineering alloys. The absence of uniformity in testing is largely responsible for the variability in mechanical property data reported in the literature.

This problem has been recognised and is being addressed by working parties in both Japan (2) and USA (3) (Table 4.1). A complete consensus has not yet been reached and further harmonisation will be necessary as the preliminary recommendations contain differences.

Table 4-1. Proposed Standards for Testing Solders

Japan	Initial casting, D, is 20 mm > test specimen D.
	Casting T, should be 100 K > liquidus, casting in < 10 s.
	After machining, anneal for 1h at 0.87 T_m (i.e. 397 and 430 K for Sn
	Pb and Sn – 3.5 Ag).
	For tensile properties, use $\dot{\varepsilon} = 2$ x $10^{-2}s^{-1}$
	measure E, 0.02 and 0.2% PS, σ_u and ε_f
	For LCF, use hourglass specimen, and radial extensometry, and
	$\dot{\varepsilon} = 10^{-3}s^{-1}$
	Define failure as '25% tensile stress drop from half life'
USA	For tensile testing,
	Specimen gl = 31.8, D = 3.81 mm
	Solder T, 100 K > liquidus
	Age at 0.67 T_m for 16h
	Allow 2 h at RT prior to testing
	Test with $\dot{\varepsilon}$ 10^{-3} to $10^{-4}s^{-1}$
	Measure E, 0.2%PS, σ_u ε_f, uniform ε_f, work hardening
	characteristics.
	LCF
	Failure defined as 50%

Perhaps of more importance is the recommendation that bulk samples should be annealed prior to testing. While this overcomes the problems of constancy of initial microstructure and its subsequent stability, the coincidence with the as-cast microstructure in an actual joint is questionable. The committees are also proposing minimum strain rates for determination of tensile properties, and alternative definitions for proof strength and fatigue life.

Comprehensive standards and guidelines regarding all aspects of the production and evaluation of electronic packages are available from the Institute for Interconnecting and Packaging Electronic Circuits (IPC) (4). These represent a valuable resource for in depth involvement.

6. MECHANICAL EVALUATION OF ELECTRONICS COMPONENTS

The application of traditional mechanical test methods to electronics requires careful attention with regard to specimen dimensions and to the magnitude of the forces involved. An actual solder joint, for example, could well be several hundred times smaller than the bulk specimen that is being used to provide data on its mechanical properties (Figure 4-6). An intermediate approach exists in the testing of *model joints*, such as pin-in-ring, or lap joints. For these, the use of conventional test machines is possible in all control modes although strain control becomes more difficult because of the dimensions of the specimen. Actual joints are not amenable to strain control testing although some quite elegant solutions to this difficulty have been produced (see Chapter 6).

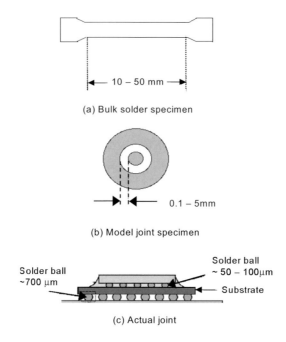

Figure 4-6. Typical size scales for bulk, model and actual joints

6.1 Evaluation of Model Joints

A model joint should simulate the most commonly encountered stress situation, be made under easily controlled conditions which are as similar as possible to those in practical soldering and, finally, give reproducible test results. Joint configurations commonly used for mechanical testing include, single lap joint, double lap joint, butt joint, ring-and-plug joint and peel tests. These are shown schematically in Figure 4-7 and considered in more detail in the following paragraphs.

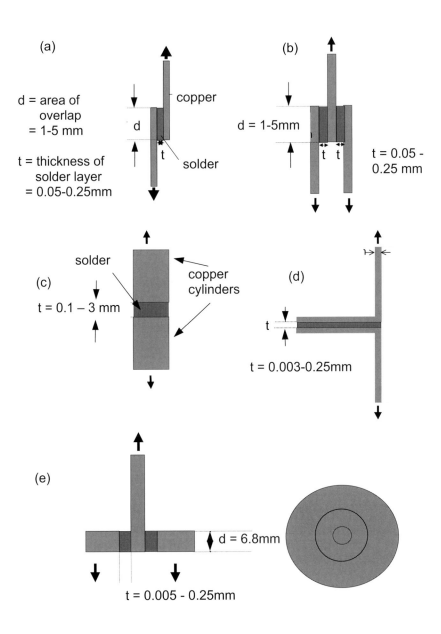

Figure 4-7. Various types of model joints

Single Lap Joint

In the single lap joint, the load is applied in a direction parallel to the joint interfaces. However, because the two component pieces cannot both lie in the axis of the load, the solder experiences rotational forces and peeling may occur.

Double Lap Joints

Here, two component pieces are soldered at either side of a third piece, and this balances out the rotational forces and eliminates bending. However, the stressing condition is not pure shear as the solder layer still experiences other stresses. The size of the whole assembly is relatively large which makes completion of the soldering process in a practicable time difficult.

Butt Joint

In this simple type of joint, the solder layer is theoretically stressed in a direction at right angles to the joint interfaces, providing a tensile stressing system. Values of shear strength may be determined from the expression introduced previously (Equation 4.1). Any non-axiality of loading leads to peeling stresses and a reduced value for strength. The risk of non-axial stressing increases with joint area. This type of joint is relatively difficult to manufacture, particularly for the larger joint gaps.

Ring-and-Pin Joint

This form of joint is often regarded as the most representative of those found in service, since it most closely gives pure shear stressing. However, the solder is not in line with the loading axis and any bending stresses due to non-axial loading are internally balanced, but not eliminated. The breaking load of the joint can be determined by pulling or pushing the pin through the ring. The latter gives slightly higher shear values due to the small radial expansion of the copper pin under compressive loading.

Peel (Chadwick) Tests

This test measures the resistance to tearing or peeling of the weakest interface in a solder joint. Narrow strips of metal are soldered parallel over a length, using aluminium foil spacers to control the joint gap. The two unsoldered ends of the component are bent away from each other to form a 'T' shape, and are pulled apart. This test effectively applies tensile stresses to the solder film at the point of peeling, and an average of the initiation and propagation stresses is taken as the peel strength.

Asymmetrical Four-point Bend (AFPB) Tests

Similar to the Iosipescu test which is commonly used for polymer and ceramic-based composites, the AFPB test produces a nearly pure shear stress state at the shear plane.

Its use for model solder joints has been demonstrated (5) and it has the merit of requiring relatively small specimens.

6.2. Evaluation of Actual Joints

Instead of defining a single failure event, mechanical assessment of damage over an entire joint may be performed. For example, a torque test or a shear displacement test on a joint provides a comparative measure of its remaining strength after thermal cycling. Indirectly, this is indicative of the damage produced and the loss of structural integrity. Any combination of board, component type, distribution and density may be evaluated mechanically provided access is available. With the recent increased interest in structural integrity in microelectronics, a new generation of test machines (*ultra high precision equipment*) is evolving, specifically for this and similar applications. The range of tests is considerable, and this is illustrated in Figure 4-8.

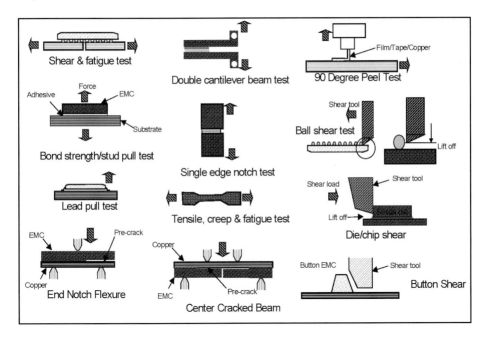

Figure 4-8. Overview of mechanical tests for microelectronic applications
(courtesy Instron Ltd)

Shear tests can be performed on complete components or single solder balls. The straddle fatigue test (6), for example, enables cyclic shear to be applied to actual flip chip interconnections by controlled displacements of the substrates (Figure 4-9). Failure is defined as a fall in load necessary to maintain the set displacement. Layer peel tests and pull-out tests on individual wires can also be carried out. Overall, much higher accuracy is necessary with respect to positioning, alignment and force measurement. For example, the resolution and accuracy requirements on displacement are 50nm and 2μm respectively. Temperature and environmental chambers may be added to the system so providing an extensive range of test conditions. The tests described above are becoming popular as indicators of damage and residual strength after thermal cycling.

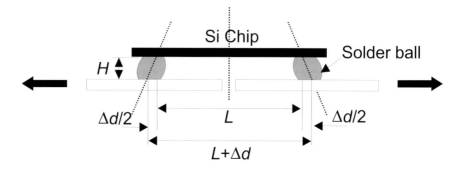

Figure 4-9. Straddle (split board) system for applying cyclic shear displacement (6)

7. EVALUATION BY THERMAL PROCESSING

The vast majority of the evaluation of interconnection and PCB reliability is achieved by thermal, rather than solely mechanical, means. As described earlier, the presence of changing temperature induces stresses and strains on a number of size scales – both microscopic and macroscopic. While uniformly shaped model joints may experience measurable, or computable, stresses and strains, actual joints do not, because of their irregular shape and the generally unknown thermal gradients that exist within them. Accurate stress-strain calculation depends upon finite element analysis, which in turn, is based on assumed values for local material properties and their variation within the joint volume.

The thermal system employed may involve thermal shock by immersion in hot and cold baths, temperature cycling in an environmental chamber or power cycling (simply turning the device on and off). Clearly, there will be significant differences in the strain rates, temperatures and temperature gradients between these methods, and such differences may affect material behaviour. For example, thermal shock is arbitrarily defined as involving temperature changes in excess of about $30^{\circ}C$ per minute (7) with failure resulting from warpage and tensile overstress. Additionally, it has been shown that power cycling is more damaging than chamber cycling for chip carrier assemblies

(8). However, it is important to note the highly specific nature of such findings. Failure generally occurs as a consequence of the balance between competing mechanisms, and a small change in either temperature range or maximum temperature can produce a change in dominant failure process.

Laboratory testing of actual components and structures for the purpose of life prediction presents a formidable challenge although detailed advice is available for standardising test procedures (4,7). There are very few instances where the component may be tested under real time conditions with realistic stress-strain-temperature histories. In most cases, *acceleration* must be introduced into the test. Further, geometrical considerations, such as shape or scale, result in complex stress and strain distributions quite specific to the component being tested. However, analysis requires generic information, which in turn necessitates evaluation of properties from quite simple specimens.

The conflict between accelerated testing of often complex, but realistic, specimens and idealised geometries from which values of stress and strain can more confidently be derived, extends to the evaluation of solder joint performance. The results of such tests, for example, with regard to the effects of various parameters, should always be assessed in terms of the match between test conditions and the real service situation. In the following sections, the principal modes of thermal and thermomechanical fatigue testing are described, and provide some perspective for the actual results that follow.

7.1 Thermal Cycling of Model Joint Specimens

This method simplifies the strain state in a joint. It involves soldering together regular shaped pieces of materials with different coefficients of thermal expansion and exposing the joint to thermal cycling. Common combinations include stainless steel-solder-aluminium, or Invar-solder-stainless steel. Polymers can also be included if they are plated. A typical specimen, comprising a central bar of Invar and outer stainless steel bars, is shown in Figure 4-10.

Figure 4-10. Model joint specimen for thermal cycling tests

Due to the regular geometry and greater uniformity of temperature, stresses and strains can be calculated or measured directly, and a greater insight into microstructural instability may be achieved. However, the technique provides no data on in-situ mechanical properties, and failure is difficult to identify apart from post-test examination.

7.2 Thermal Cycling of Actual Components or Boards

The benefits of this approach are that it utilises actual specimens or assemblies, and that deterioration can be monitored continuously, for example, by electrical means. Numerous specimens can be tested simultaneously which facilitates statistical analysis. Acceleration of testing (typically 100 to 1000 times) is achieved by a substantial reduction in the dwell duration in the temperature-time profile, and raising heating and cooling rates. This type of evaluation is empirical, and provides highly specific information on the component – (PCB) – temperature profile combination being tested. Table 4.2 illustrates some of the possible variables that could be considered, and demonstrates the enormous difficulty of the sole reliance on this approach, even for apparently minor alterations in operating conditions, such as changing the temperature range or heating and cooling rates.

Table 4-2. Summary of Possible Variations in Thermal Cycling of PCBs

Parameter	Variables
Board	- dimensions, materials
Component	- type, size, number, arrangement
Service	- T_{max}, T_{min}, t_{max}, t_{min}
	rates of heating and cooling
Interconnection	process history, microstructure
What if service conditions change?	

Extrapolation to real time conditions is hazardous, and many of the reasons for uncertainty have yet to be fully appreciated. For example, a faster strain rate may be associated with higher strength. However, as strength increases, the propensity for sudden catastrophic fracture is also increased. At lower strain rates, deformation might be accommodated within the materials grains by anelasticity and plasticity, while at very low strain rates, grain boundary processes are likely to dominate. In terms of effect on structural integrity, this simple example indicates that strain rate can have different, and even opposite, effects according to its value and the controlling deformation mechanism (Figure 4-11). Similarly, when acceleration is achieved by curtailing dwell periods, stress relaxation is restricted which may promote additional creep deformation in unconstrained regions of a joint. Within constrained regions, the initial and rapid stress relaxation converts to matrix strain, and is not particularly deleterious. However, if the strain rates during stress relaxation are sufficiently low to promote grain boundary damage, this creep damage is enhanced and endurance reduced.

Until reliable deformation or fracture mechanism maps are available for materials in interconnections, the empirical approach will be required.

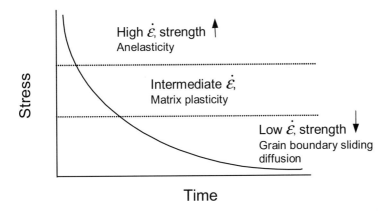

Figure 4-11. Schematic illustration of the effect of strain rate on damage modes during stress relaxation

7.3 Thermomechanical Fatigue of Model Joint Specimens

Idealised joints are subjected to externally applied cyclic strains and temperatures. In essence, this is the same as isothermal strain controlled fatigue – with an additional temperature variation that may be in-phase (IP) or out-of-phase (OP) with the applied strain. Figure 4-12 illustrates schematically the various modes of testing. Heating is generally achieved either by resistance coils around the specimen, or by RF induction, and cooling by forced air or liquid coolants.

Monitoring stress response enables mechanical properties to be determined during testing although the correlation with actual joint behaviour has yet to be established. Since this form of test is quite complex and expensive, interest remains in the extent to which isothermal strain controlled fatigue is an indicator of TMF behaviour.

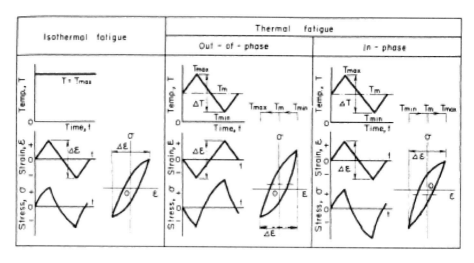

Figure 4-12. Various modes of testing in isothermal and thermomechanical fatigue (9)

7.4 Thermomechanical Fatigue of Actual Solder Joints

Multiple soldered joints between panels may be cycled in shear under *displacement* control with external air heating and cooling. Continuity monitoring enables failure in a particular joint to be identified. While this method meets the requirements of scale, uniformity of strain and a realistic failure criterion, it is quite complex and time consuming and requires highly sensitive strain resolution.

7.5 Mechanical v Thermal Testing

A point which manufacturers identify as a strong advantage of mechanical, rather than thermal, testing is the substantially shorter duration of the former. For example, an isothermal mechanical fatigue test over the same strain range as an equivalent thermal cycling test may be several hundred times shorter (Figure 4-13). However, this benefit merits close scrutiny. The problem of selecting which isothermal temperature and strain rate to employ is compounded by the variation in influential mechanical properties with both temperature and strain rate.

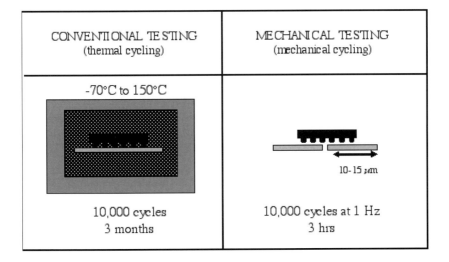

Figure 4-13. Comparison between thermal and mechanical cycling experiments

8. CONCLUSIONS

The determination of mechanical properties from bulk samples of regular dimensions is well established and as reliable as the quality of the sample. Direct measurements of stress and strain from the gauge length of the specimen avoid uncertainties about test machine flexibility and stress or strain distribution around the specimen shoulders. In this regard, they are preferred for the evaluation of *intrinsic material properties.*

Soldered interconnections may be 100-1000 times smaller than conventional bulk specimens. Even model joints, such as lap or pin-in-ring, are usually an order of magnitude smaller. Mechanical testing of such samples, therefore, requires greater sensitivity in terms of force and displacement. A new generation of test equipment is emerging specifically for this market. Accelerated thermal cycling provides valuable information specific to the system under test, although its empirical nature is a significant shortcoming. It is becoming increasingly recognised that mechanical property information on components and interconnections has a valuable contribution to make to reliable life prediction and design.

9. REFERENCES

1   HD Solomon, Electronic Packaging: Materials and Processes, ASM, 1986, 22.
2   M Sakane, H Nose, M Kitano, H Takahashi, M Mukai and Y Tsukada, Dec, 2001, Tenth Int. Conf. on Fracture, ICF 10, Hawaii.
3   T A Siewert and C Handwerker, Test Procedures for Developing Solder Data, 2002, NIST, Special Publication 960 – 8.
4   For example, IPC-9701, Performance Test Methods and Qualification Requirements for Surface Mount Attachments, IPC, Jan, 2002.

5   O Unal, DJ Barnard and IE Anderson, Scripta Materialia, 1999, 40, 271.
6   HD Solomon, IEEE Trans Components, Hybrids and Manufacturing Technology, 1989, 12, 473.
7   IPC-5M-785, Guidelines for Accelerated Reliability Testing of Surface Mount Solder Attachments, p5, 1982.
8   J Lynch and A Boelti, Thermal Stress and Strain in Microelectronics Packaging (Ed J H Lau), Van Nostrand Reinhold, (New York), 1993, 579.
9   A Nitta and K Kuwabara, High Temperature Creep-Fatigue, (Eds R Ohtani, M Ohnami and T Inoue), Elsevier Applied Science, (London and New York), 1988, 203.

CHAPTER 5

MECHANICAL PROPERTIES OF BULK SOLDERS

1. INTRODUCTION

Having examined the basic principles of the processes that govern structural integrity and performance, attention is now turned to actual values of the relevant properties. The purpose of this is twofold:
1. to provide a quantitative appreciation of the properties involved, in particular, the relative merits of lead-bearing and lead-free solders;
2. to demonstrate the unusual characteristics of solders, amongst common engineering alloys, in the degree of variability and uncertainty that exists in the published values for basic mechanical properties.

In this context, *variability* exists due to the absence or disregard of standards for testing. For example, details regarding prior thermal history, surface condition, strain rate or mode of control may vary between laboratories, or are not cited in the published literature. *Uncertainty* or inconsistency arises because of the high homologous temperatures at which solders normally operate. The change in the unstable microstructure is inevitably mirrored by changes in mechanical properties. So, any particular property measured at one time under a well-defined set of conditions may have changed when re-assessed subsequently. Naturally, the degree of instability is dependent upon the prior history of the sample and compounds the variability factor described above. The key point of caution here is that any quoted property value may be significantly affected by the material condition and the testing parameters, and might not remain constant thereafter. Unlike most other engineering alloys, which generally operate at much lower homologous temperatures, there are few fixed property values.

Given this large number of possibilities, it is not surprising that few sets of results are directly comparable. The exception is when data have been produced by the same workers in the same laboratory. Even in the same company, variations in prior thermal history may be found! So while a considerable amount of data may be available, its value is sometimes limited. All published data require close scrutiny.

In addition to impaired comparability, an equally important feature is the match between the conditions under which the properties were evaluated and those likely to be encountered in service. Improving that coincidence produces data of most relevance to modelling.

The present chapter examines the mechanical properties of solders in their bulk form, while the subsequent chapter reviews the mechanical performance of model and actual joints.

2. MONOTONIC PROPERTIES

As described previously, these relate to behaviour under a continuously increasing stress. For convenience and to reflect the amount of data available, emphasis is placed upon tensile behaviour.

2.1 Modulus Values of Solder Alloys

The modulus of elasticity (E or G, in tension or shear) is determined by the forces between adjacent atoms and is often described in textbooks as a material constant. It diminishes with increasing temperature. For the vast majority of engineering materials (steels, copper alloys, ceramics etc.) published figures for moduli confirm their constant values. However, published modulus values for solders (1) exhibit an enormous variation (Table 5.1).

Table 5-1. Reported Values of Young's Modulus for Solder Alloys at Room Temperature

Alloy	Modulus (GPa)	Alloy	Modulus (GPa)
Sn	58.2	Sn-37Pb	5.7
Pb	14	"	20.5
Pb	45.8	"	35
Sn-40Pb	38.6 ± 4	Sn-3.5Ag	50
"	16	"	56
"	15	Sn-5Sb	50
"	30	"	58
"	35	Sn-58Bi	42
Sn-37Pb	32	Sn-52In	23.6
"	32-34	Sn-2.5Ag-0.5Sb	29.7
"	40		

The significance of the modulus in the interpretation and analysis of mechanical performance is that it determines the strain per unit of applied stress under elastic conditions. Similarly, in conjunction with the yield stress, σ_y, it governs the limit of elastic strain that occurs prior to the onset of plasticity. For example, during thermal cycling, which is strain controlled, the influential plastic strain range, $\Delta\varepsilon_p$, may be determined from equations (2.1) and (3.8).

$$\Delta\varepsilon_p = \Delta\varepsilon_t - \Delta\varepsilon_e \text{ and } \Delta\varepsilon_e = \Delta\sigma_y / E \qquad (5.1)$$

where $\Delta\varepsilon_t$ and $\Delta\varepsilon_e$ are the total and elastic strain ranges respectively.

Fortunately, in this case, the yield strength of solders is extremely small, and in most situations.

$$\Delta\varepsilon_p \approx \Delta\varepsilon_t \qquad (5.2)$$

Conversely, the lower the strains experienced, the more significant the modulus value becomes in the determination of the plastic strain range.

Modulus is also described in many textbooks as *structure insensitive*, so apart from displaying a change with temperature, the value for tin and tin-based alloys should be essentially constant, irrespective of their prior history. However, inspection of the reported values for Young's modulus for solder alloys (Table 5.1) reveals a wide variation even for the same alloy. Most values reported have been determined from the slope for the 'elastic' region of the stress-strain graph, and it is probable that strain-rate effects have had a significant influence in the observed scatter. High sensitivity stress-strain measurements (2,3) indicate a distinct difference in initial slope, and hence in deduced value of modulus, as the strain rate changes (Figure 5-1). The value of Young's modulus at room temperature for tin–37 lead determined from dynamic mechanical analysis and acoustic-pulse methods is about 30 GPa. (4). This was about an order of magnitude greater than that determined on the same material, in the same condition, from a mechanical test (Figures 5.2 and 5.3). Given the fact that most lead-free solders are tin-based alloys, it is appropriate to conclude that the moduli of lead-free solders are broadly similar to tin-lead alloys.

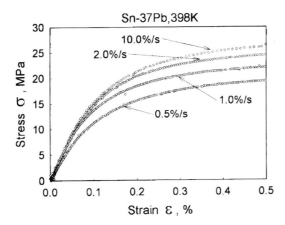

Figure 5-1. Influence of strain rate on the initial stages of tensile deformation (3)

Accurate modelling of thermomechanical fatigue requires knowledge of the material properties over the temperature range covered. These data are rarely available, and it has been common practice to use the room temperature or the 'average' values of properties at the temperature extremes.

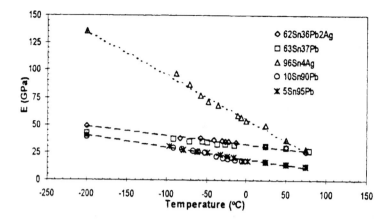

Figure 5-2. Effect of temperature on elastic modulus determined from acoustic pulse tests (4)

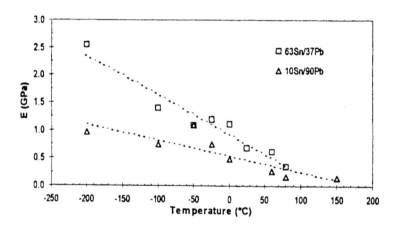

Figure 5-3. Effect of temperature on modulus determined from tensile tests (4)

Jones et al (4) have determined linear relationships between modulus and temperature over the range –200°C to 100°C for several lead-containing solders, and tin-3.5 silver (Figures 5.2 and 5.3). With E in GPa and T in °C, the empirical relationships were:

Sn-37Pb,	$E = 31.5 - 0.057\ T$	(5.3)
Sn-95Pb,	$E = 19.5 - 0.107\ T$	(5.4)
Sn-4Ag,	$E = 55.4 - 0.41\ T$	(5.5)

All these alloys were in the as-cast into water-cooled condition.

2.2 Stress-Strain Response

The stress-strain response of solders is typical of that of ductile metals and alloys that exhibit a small and indistinct yield point. Conventional proof stress values, say 0.2 per cent, are generally well into the plasticity domain (Figure 5-1), and while it is possible to measure them, comparability between laboratories is more unreliable due to inconsistencies in their measurement. It is also likely that creep is occurring during the early stages of loading, which further masks the onset of yield. In contrast, the tensile strength is more directly measurable, and is most often used here as an indicator of strength.

2.2.1 Strength of Solders

Solders exhibit the normal trend of a gradual reduction in strength with increasing temperature (Figure 5-4). Lead-rich, solid solution, alloys tend to be weaker but more ductile, in terms of uniform elongation, than eutectic solders. The latter are susceptible to low ductility (approximately 5 per cent) fracture at temperatures below about -150°C. Generally, strength is an indicator of resistance to high cycle (low stress) fatigue, whereas low cycle fatigue is more influenced by ductility. Thermomechanical fatigue is usually a low cycle, high plasticity process. Jones et al (4) have developed empirical expressions for the effect of temperature on strength.

Sn-37Pb   (T > - 50°C)

$$\sigma_u = 49 - 0.32 \, T \tag{5.6}$$

Sn-95Pb   (-200°C < T < 150°C)

$$\sigma_u = 32 - 0.17 \, T \tag{5.7}$$

Sn-3.5Ag   (T > - 50°C)

$$\sigma_u = 62 - 0.34 \, T \tag{5.8}$$

These data were generated at strain rates of $2 \times 10^{-2} \, s^{-1}$ on rapidly cooled, fine-grained, specimens.

2.3 Examination of Data

To illustrate the previous points and to avoid the problems in comparing data from different laboratories, the following section presents findings of the Solder Research Group at the Open University (5). These measurements were made on cylindrical bulk specimens which, unless otherwise stated, were in the as-cast and rapidly cooled condition. The ensuing microstructure most closely resembles that found in actual joints. Specimens were stored in a freezer at −19°C after casting and received no further treatment prior to testing.

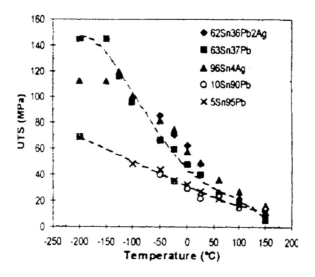

Figure 5-4. Influence of temperature on strength (4)

2.3.1 Influence of Temperature

For the eutectic Sn-37Pb alloy, strength is approximately halved as the temperature is increased from –10 to 75°C, (Figure 5-5) (6). At slower strain rates, the effect of temperature is enhanced and an decrease in strength of around four times is observed at 10^{-3} s^{-1} for the same temperature change. The Sn-3.5Ag and Sn-0.5Cu alloys exhibit similar behaviour, but their strength is less sensitive to temperature in the range examined. Proof strength (0.2 per cent) values tend to follow the same pattern.

Ductility, as determined by either elongation to failure or reduction of area, generally increases with increasing temperature, although it is less affected than strength (Figure 5-6). Again, the lead-containing solder was most sensitive to temperature.

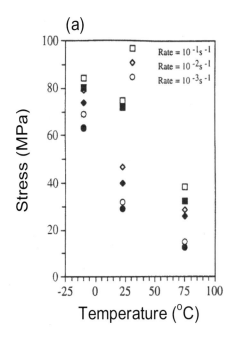

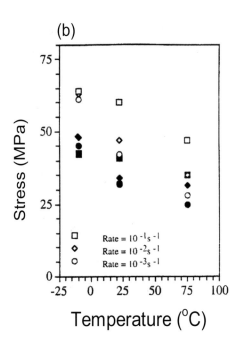

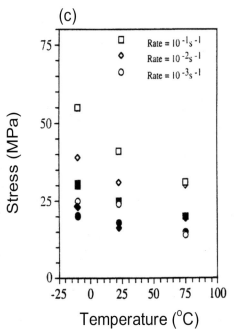

Figure 5-5. The influence of temperature on the strength of (a) Sn-37Pb, (b) Sn-3.5Ag and (c) Sn-0.5Cu at various strain rates. Open points: Tensile Strength, Closed points: proof stress (0.2%)(6)

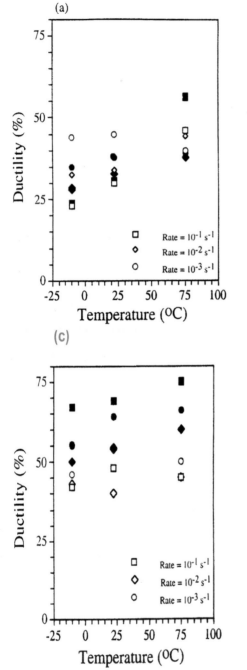

(a)

(b)

(c)

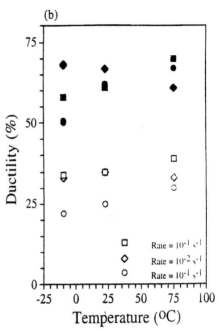

Figure 5-6. *The influence of temperature on ductility at a different strain rates for: (a) Sn-37Pb, (b) Sn-3.5Ag and (c) Sn-0.5Cu. Open points: Strain to failure, Closed points: Reduction of area (6)*

2.3.2 The Effect of Strain Rate

Within the range 10^{-3} to 10^{-1} s^{-1}, tensile strength increases substantially with increasing strain rate (Figure 5-7). Typically, a doubling in strength is observed, although the effect is usually less at $-10°C$, when the Sn-3.5Ag alloy is hardly affected by strain rate. The progressive loss of strength with diminishing strain rate does not extend to very low strain rate regimes (10^{-6} s^{-1}). Strain rate affects the ductility of the alloys differently. There is little change for the Sn-0.5Cu alloy; the Sn-3.5Ag solder shows an increase in ductility at higher strain rates, and opposing trends are observed in the Sn-37Pb alloy, according to the temperature.

Figure 5-7. The effect of strain rate on strength at $-10°C$ for various solder alloys

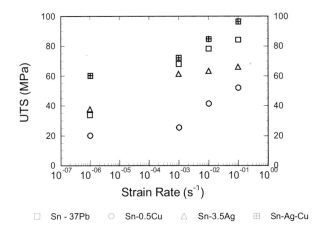

2.3.3 Comparative Behaviour of Lead-Containing and Lead-Free Alloys

The wide and variable effects of temperature and strain rate require that even ranking of alloys should be made under clearly defined conditions. This is demonstrated in Figures 5.7 and 5.8 that include data for the ternary Sn-3.8Ag-0.7Cu. At $-10°C$, the ranking in the alloys changes with strain rate. In contrast, at $75°C$, the superiority of the silver-containing alloys is retained over all the strain rates examined, although the strengths of the Sn-0.5Cu and Sn-37Pb alloys tend to converge with decreasing strain rate. The stronger alloys are slightly less ductile.

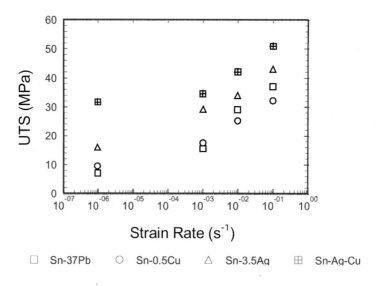

Figure 5-8. *The effect of strain rate on strength at 75°C, for various solder alloys*

2.4 Summary of Monotonic Behaviour

There is much diversity in the testing procedures used for determining the monotonic properties of solders. Considerable scatter exists in the reported values for elementary properties, such as elastic modulus and tensile strength. Both temperature and strain rate are highly influential, so that not only absolute values of properties change, but also their ranking may be affected. Selection of alloys for design and modelling should take account of the potential conditions in service. The silver-containing alloys are generally stronger than the tin-lead eutectic, and the Sn-0.5Cu alloy has a similar strength to Sn-37Pb.

3. CREEP BEHAVIOUR

The concept of time-dependent (creep) response of materials was introduced in Chapter 3, where graphical representation of creep data was reviewed. In addition, popular empirical equations, used to predict creep lifetime, were considered. In service, creep may appear as deformation, under stress-controlled conditions, or as stress relaxation when strain-limited conditions exist. The aim of this section is to review available creep data for lead-free solders using the eutectic tin-lead as the comparator alloy. The problems of comparability of data have been introduced previously. Creep is particularly vulnerable to such difficulties, since as a generally low stress phenomenon, it is highly sensitive to both microstructure and its stability.

3.1 Creep of Eutectic Tin-Lead Alloys

A significant proportion of the available data is of limited value to the designer because it relates to high and unrepresentative stress levels (low rupture times). Such information most probably describes creep mechanisms that do not operate in the low stress regime, and as discussed previously, extrapolation to longer life times is unsafe.

The scatter in data for creep behaviour is much greater than that for other modes of deformation, such as monotonic straining or fatigue. This variability is demonstrated in Figure 5-9 which shows a graph of minimum strain, $\dot{\varepsilon}_m$, v applied stress at two common test temperatures, room temperature and 75 (or 80) °C.

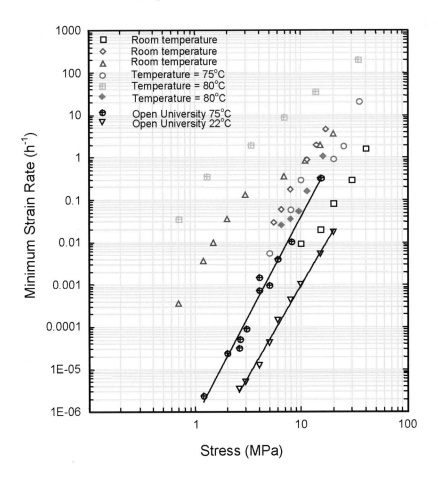

Figure 5-9. *Example of data scatter in broadly similar creep tests on Sn-37Pb solder*

This type of plot is preferred, rather than that of a graph of stress against time to rupture, since it eliminates the potentially confusing effects of ductility. The salient feature in the present context is the dilemma faced by the designer or stress analyst in deciding which data to choose for their calculations.

Figure 5-9 indicates that for a high stress level of say, 10 MPa (roughly equivalent to lifetimes of 10h at room temperatures and 1hr at 75°C) the scatter in measured minimum strain rate values is 2-3 orders of magnitude. At lower, and more realistic, stress levels (e.g. 3 MPa) the scatter appears even greater. Table 5.2 provides an explanation for this apparent variability in creep behaviour. The substantial differences in activation energy and stress exponent confirm that different investigators were examining different regimes of creep. Given that many more data sets exist for the Sn-37Pb alloy, yet relatively few are available for the lead-free systems, the benefit of making global comparisons from the literature between the eutectic and lead-free alloys is uncertain.

Table 5-2. Variability of Activation Energy and Stress Exponents for Tin-based Solders

Alloy	*Activation Energy* KJ/mol	*Stress exponent 'n'*	
		Single reported value	*Minimum reported and maximum reported*
Sn-37Pb	56.9	3.04	10.75 – 15.79
			7.27 – 12.49
			4.27 – 9.02
			3.61 – 6.71
			3.19 – 6.22
			3.09 – 5.0
	61.4	None reported	1.9 – 5.9
	61.4	1.88	1.89 – 2.61
	54.7	2.03	2.05 – 2.91
			2.09 – 4.74
	91.8		2.4 – 5.7
	95.8		3.0 – 7.0
	none reported		3.1 – 7.3
	102		2.0 – 7.1
	93.8		1.7 – 11.1
	140	3.4	
Sn-3.5Ag	84	5.5	
	54.6	5.0	
	142.8	10	
	111.3	12	
	81.6	5.5	
	60.7	5.0	
Sn-Ag-Cu	61	13	
	99	14	

3.1.1 Factors Influencing Creep of the Eutectic Tin-Lead Alloy

In the following section, the parameters (such as temperature, microstructure and prior history) which may influence creep behaviour are examined.

The range of temperature over which solders may have to operate is extremely wide (say -55°C to 160°C). Reliable modelling requires data for temperatures within this range, although few exist at the extremes. Figure 5-10 shows the stress v time to rupture plot for the Sn-37Pb alloy as a function of test temperature (7). Contrary to some assumptions, it is clear that creep occurs at temperatures as low as –50°C. For an applied stress of about 15MPa, the ratio of the creep lives at 75, 25 and –40°C is approximately 1:10:1000. The minimum strain rate ratios are also similar. Creep ductility is significantly influenced by temperature. At 75, 25 and –40°C, ductility ratios are 6:3:1. In addition, there are combinations of stress, temperature and microstructure where the possibility for superplastic deformation exists.

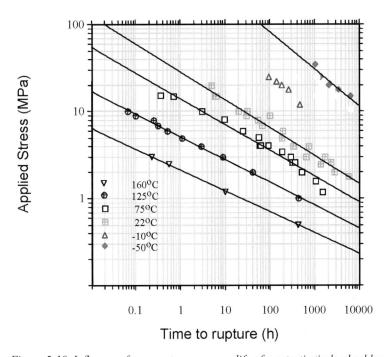

Figure 5-10. Influence of temperature on creep life of a eutectic tin-lead solder (7)

Apart from temperature, the outcomes of other parameters on creep behaviour are less clear cut and are often dependent upon the microstructure of the solder. For example, initial cooling rate has a profound influence on the microstructure of eutectic lead-tin solders (Figure 5-11) but the resulting creep performance at 75°C is quite similar for the air, water and furnace cooled conditions (8, 9). In contrast, prior ageing (100h at 75, 100 or 130°C) is increasingly deleterious as the rate of initial cooling increases. Reductions in life exceeding an order of magnitude were observed for the water-quenched condition (Figure 5-12). A prior strain of 5 per cent does not affect creep life in this condition, but causes a reduction in time to rupture of 2-3 times for the slower cooled states. Creep strains to failure are enhanced by up to five times by ageing and prior straining of the air cooled and water quenched alloys, whereas creep ductility in the furnace cooled condition is generally unaffected by prior ageing (9). Prior cyclic strain causes softening in many solders, and this reduces subsequent creep life by about a factor of three, with ductility little affected (10). The complexity of these findings indicates the care required in designing and interpreting tests on actual joint samples.

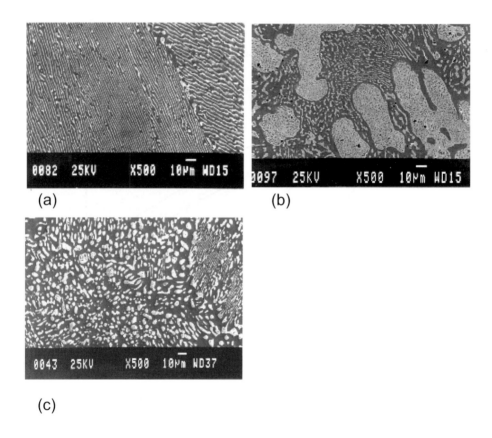

Figure 5-11. Range of microstructures produced by different cooling rates in Sn-37Pb
(a) water quenched, (b) air cooled, (c) furnace cooled.

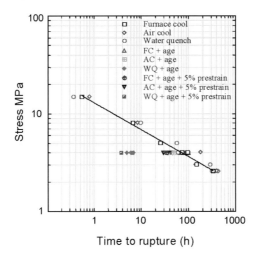

Figure 5-12 Effect of prior ageing and pre-strain on creep of Sn-37Pb solder

3.2 Creep Performance of Lead-free Solders

Figure 5-13 presents the stress v log (time to rupture) data at 75°C for four alloys (11, 12). The maximum rupture time generally exceeds 1000h, and in some cases approaches 10,000h. The range of behaviour encompassed by the alloys is substantial, to the extent that it is sometimes difficult to compare them in terms of applied stress. For creep lives below 1000h, the ternary Sn-Ag-Cu alloy is markedly superior to the Sn-Ag binary, which in turn, is substantially better than either the Sn-Cu or the Sn-37Pb binaries. For example, with an applied stress of 15MPa, the rupture life of the ternary is about 100 times greater than that of the silver-containing binary and 1000 times higher than Sn-Cu or Sn-Pb. For lower stress levels and lives in excess of 1000h, there are indications of changes in slope that will affect these comparisons. In terms of creep strength at 1000h, the values are approximately 17 MPa (Sn-Ag-Cu), 11 MPa (Sn-Ag), 3 MPa (Sn-Cu) and 2 MPa (Sn-Pb).

No clear trend exists between the creep strain to failure, ε_f, and either applied stress or time to rupture. Both silver-containing alloys are generally less ductile than the Sn-Cu or the Sn-37Pb alloys. The vast majority of strains to failure fall between 10 and 50 per cent.

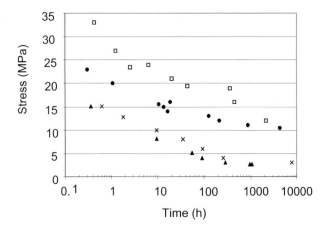

Figure 5-13. Stress v time to rupture for lead-free alloys at 75°C (11)

Figure 5-14 shows a graph of minimum strain rate versus applied stress. Comparison between the alloys is impaired because of the differences in strength and the inflection in the ternary alloy data. However, with an applied stress of 12 MPa, the estimated minimum strain rate values are 3×10^{-6} h^{-1} (Sn-Ag-Cu), and 2×10^{-2} h^{-1} (Sn-Pb) respectively – a range of some four orders of magnitude. The relationship between minimum creep rate and applied stress for solders has been described as sigmoidal (13), and there are indications of such transitions. The stress exponent of Sn-Ag was found to be approximately 12; that of Sn-Cu was 5 and that of Sn-Ag-Cu was 17. However, the stress exponent for Sn-Pb showed a transition from 3 at low stresses to 7 at high stress levels.

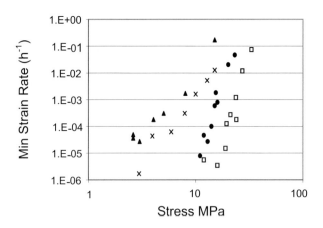

Figure 5-14. Minimum strain rate v applied stress at 75°C (11)

Minimum strain rate and ductility characteristics may often be reconciled using the Monkman-Grant expression

$$\dot{\varepsilon}_m t_r = C \qquad (5.9)$$

Figure 5-15 demonstrates the linear trend exhibited by all data.

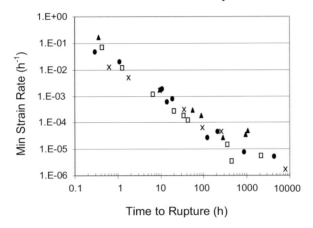

Figure 5-15. Monkman-Grant plot at 75 °C

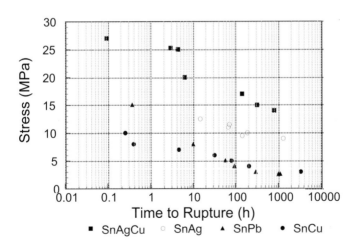

Figure 5-16. Creep performance at a constant homologous temperature (T_h = 0.76)(12)

Differences in creep resistance may be attributable to variations in melting point. Intrinsic differences between alloys can be revealed only by testing at similar homologous temperatures, T_h (Recall that T_h is the ratio of the test temperature to the melting point in Kelvin). Tests performed at $T_h = 0.76$, with the following test temperatures; Sn-Cu (107°C), Sn-Ag (102°C) and Sn-Ag-Cu (99°C) reveal the inherent superiority of the silver-containing alloys (5) (Figure 5-16).

The efficacy of the Monkman-Grant equation is dependent upon the dominance, or otherwise, of the steady state creep phase. Figure 5-17 shows that this stage is usually less than one third of the entire creep life for the solder alloys. Given the somewhat arbitrary definition of the extent of steady state creep as the time fraction during which the creep rate is within 10 per cent of its minimum value, it is evident that secondary creep has a similar significance for all the alloys examined.

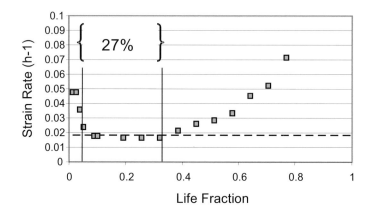

Figure 5-17. Variation of incremental strain rate with life fraction, showing extent of 'steady state' creep (12)

The influence of an intermediate ageing treatment (1h at 100°C) between casting and creep testing has been shown to reduce the high stress creep life of Sn-3.5Ag by a factor of around three (14) with both minimum creep rate and ductility being increased (Figure 5-18).   In this case, the dominant creep mechanism was unchanged.

This finding reflects the differences arising from various standards described in Chapter 4. In particular, incorporation of a stabilisation anneal prior to testing may affect all mechanical properties. The Japanese Society for Materials Science have produced a systematic compendium of tensile and fatigue properties for Sn-37Pb and Sn-3.5Ag alloys after such an anneal (15) and the study is to be extended.

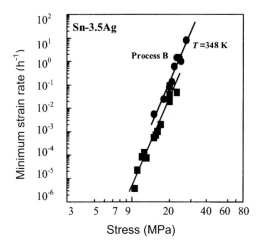

Figure 5-18. Effect of ageing (1hr at 100°C) on minimum creep rates in an Sn-3.5Ag alloy (14)

3.3 Summary of Creep Behaviour

Creep is the most sensitive deformation mode, both to the testing conditions and to the microstructure. Substantial scatter is observed in the data for solders, which creep at temperatures down to –50°C. The silver-containing solder alloys exhibit improved creep lives over Sn-37Pb and Sn-0.5Cu alloys, even when testing at the same homologous temperature. Ductility is lower in these alloys, but not restrictively so. Intermediate ageing after casting impairs creep behaviour of both types of alloy, although the extent of this depends upon the cooling rate after casting.

4. FATIGUE BEHAVIOUR

Failure under high strain fatigue, induced by fluctuating temperatures which determine the strain limits, has already been identified as a principal cause of interconnection failure in service. The present section examines the fatigue properties of solder alloys in the context of utilising design approaches based upon physical metallurgy and failure mechanisms.

4.1 'Stress' and 'Strain' Fatigue

A misconception of many engineers and materials scientists is that 'stress' and 'strain' are interchangeable, and this arises because of the tendency to think only 'elastically'. This can cause confusion in interpreting fatigue behaviour. Due to their low yield strengths, solders generally experience high strain fatigue

($\Delta\varepsilon_p >> \Delta\varepsilon_e$). As a general rule, the low strain (low stress) response of a material is governed principally by its strength because cycling is largely elastic, and stress and strain are interchangeable. In contrast, cycling between high strain limits imposes a permanent strain on to the material without any reference to the stress level required to achieve the strain limit. Under such circumstances fatigue endurance is determined mainly by material ductility. When following this argument to investigate what external factors may influence fatigue behaviour, it is essential to be aware of which particular fatigue domain is being considered. So, values of strength, and factors that affect it, broadly determine high cycle fatigue endurance, while ductility and its controlling parameters govern low cycle fatigue life. Inspection of the mechanical properties of common engineering alloys indicates a much wider variation in strength than in ductility. Therefore, the low cycle fatigue behaviour of most alloys is neither diverse nor particularly sensitive to microstructural features that may profoundly affect strength and high cycle fatigue behaviour. Figure 5-19 demonstrates this similarity in endurance for solders and other alloys at broadly similar homologous temperatures (16).

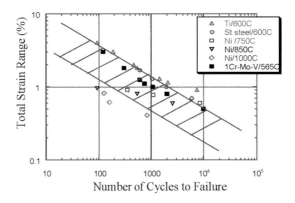

Figure 5-19. Plot of strain range v cycles to failure for a range of engineering alloys. The hatched domain includes solders at RT and 75°C (16)

The situation with a soldered joint may be best described as one in which the solder experiences low-cycle fatigue under limits of strain. An obvious question arising from this is the extent to which isothermal, strain controlled, fatigue data may be a useful indicator of TMF performance. A recent strain analysis by Raeder et al (17) acknowledges the fact that the real situation in an actual joint is only partially constrained. Consequently, both stress and strain in a soldered joint may change during thermal cycling as the interaction with the entire assembly (substrate, component and possibly leads) develops. Use of a reference assembly stiffness enables determination of more appropriate stress and strain levels in the solder, and incorporation of an 'effective plastic strain range' (the average of the stable and the failure ranges) facilitates the application of the Coffin-Manson expression.

4.2 Fatigue Performance During Continuous Cycling

The isothermal, strain controlled, fatigue behaviour of traditional lead-containing solder alloys, particularly the eutectic, has been extensively examined and reviewed (1). While many exhibit cyclic softening, a few show cyclic hardening (1). Most data give a good fit to the Coffin-Manson expression although some examples of bilinearity in this plot are reported (18). The majority of the information available relates to a fairly narrow range of test conditions, so it is not easy to develop an overview of behaviour over the potentially broad environment likely to be encountered in service.

This shortcoming has been addressed by Shi et al (19) who have examined low cycle fatigue in a Sn-37Pb alloy over the temperature range –40 to 150°C, and the frequency range 10^{-4} to 1 Hz. The fully reversed, total strain, level reached 50 per cent, and the specimens were machined from cast bars prior to annealing at 60°C for 24h in a nitrogen atmosphere. Little scatter was found in repeated tests and a good fit was obtained with the Coffin-Manson expression over the entire strain spectrum examined. The fatigue life decreased gradually with increasing temperature (Figure 5-20), and the values of the Coffin-Manson constants, C and m, also changed. (m was slightly affected, changing from 0.763 at –40°C to 0.69 at 150°C, whereas the value of C fell from 2.75 at –40°C to 0.98 at 150°C). As the frequency of cycling decreases, the fatigue life falls. For frequencies above around 10^{-3} Hz, a weak dependency is observed, but below this, the life falls substantially with reduction in frequency (Figure 5-21). In other high temperature alloys, such as stainless steels, fatigue endurance again becomes insensitive to frequency at very low frequencies. It is possible that this might occur in the solder alloy at frequencies below those examined. The value of C from the Coffin-Manson model decreases significantly with decreasing frequency, whereas the value of m shows little frequency dependence.

This comprehensive investigation (19) provides a valuable insight into the volume of information necessary for reliable modelling even for the relatively simple case of continuous cycling. However, the utilisation of frequency, rather than strain rate does impose a potential limitation. With a fixed frequency, the variation in strain rate is proportional to the limits of strain examined, which in this study was a factor of 50. Such a variation can significantly affect mechanical behaviour.

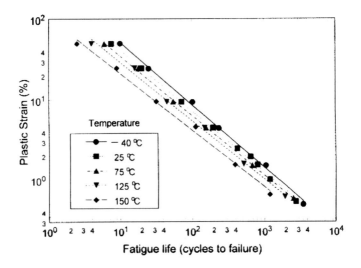

Figure 5-20. Effect of temperature on fatigue life for Sn-37Pb (19)

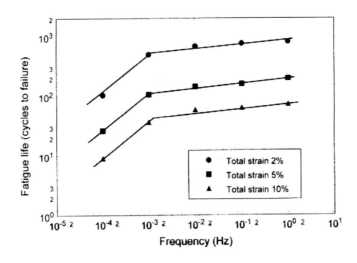

Figure 5-21. Influence of frequency on fatigue life for Sn-37Pb (19)

Nose et al (20) have investigated strain rate effects on the fully-reversed torsional fatigue of eutectic tin-lead at 40°C. Specimens were annealed at 100°C after machining into shape. Employing the Von Mises inelastic strain range

$$\Delta\varepsilon_{in} = \Delta\gamma\sqrt{3} \qquad (5.10)$$

in the Coffin-Manson plot, they found that strain rate had little effect on fatigue life during symmetrical cycling (i.e. when the strain rate is constant during entire cycle). Endurances for a strain rate difference of two orders of magnitude fell within a factor or two scatter band (Figure 5-22). However, asymmetric cycling produced substantial reductions in endurance up to almost fifty times for a $5 \times 10^{-5} - 5 \times 10^{-3}$ s^{-1} cycle profile.

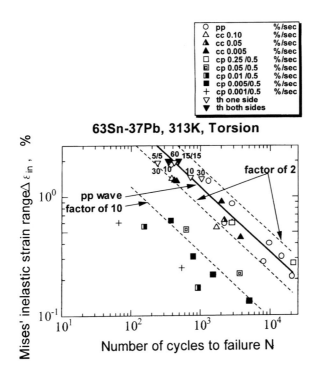

Figure 5-22. Effect of different cycle profiles on fatigue endurance for Sn-37Pb (20)

A similar insensitivity to strain rate during symmetrical cycling has been reported for lead-free alloys (21). A slow-fast cycle profile caused substantial reductions in endurance for the Sn-3.5Ag-X series of alloys (X is bismuth or copper), and the converse profile, fast-slow, was damaging to Sn-3.5Ag and Sn-3.5Ag-1Cu, but the endurance of the bismuth-containing alloy was unaffected.

The influence on the fatigue performance of Sn-3.5Ag of minor alloy additions has been studied in depth by Kariya and Otsuka (22,23). Their specimens were heat treated at 100°C for one hour after machining, and cycled in fluctuating tension at room temperature. The endurance of the binary Sn-3.5Ag alloy was about ten times that of the eutectic Sn-37Pb alloy. Increasing amounts of bismuth reduced fatigue life, and above two per cent it was inferior to the Sn-37Pb alloy (Figure 5-23). It was proposed that fatigue was influenced by tensile ductility, D, as determined by reduction in area, RA, and that the Coffin-Manson expression might be modified to

$$\left(\Delta\varepsilon_p / 2D\right)N_f^m = C \tag{5.11}$$

D is given by ln $\{100/(100 - RA)\}$

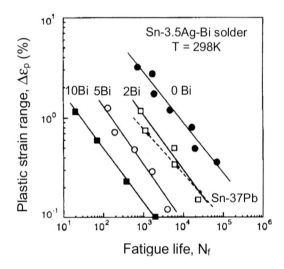

Figure 5-23. Influence on small additions of bismuth on the fatigue performance of Sn-3.5Ag (23)

In contrast to bismuth, additions of copper, zinc and indium only slightly impaired the fatigue performance of the Sn-3.5Ag alloy. All alloys studied remained superior to the Sn-37Pb eutectic.

4.3 The Effects of Dwell Periods

During service, most components experience periods of constant temperature during which the induced strains are fairly constant. This is usually simulated in isothermal fatigue testing by the insertion of a dwell at one, or both, strain limits. Substantial reductions (in excess of an order of magnitude) in the number of cycles to failure have been reported for many high-temperature engineering alloys (15). Similar observations have been made for solders (18) experiencing tensile-only dwells under some test conditions, but not for others. A saturation of the dwell effect after hold periods of 100s has sometimes been observed (24), and the damage process has been identified as mechanical rather than environmental. Balanced dwells in tension and compression have little effect on endurance (25). Contrasting results have been obtained recently by Nose et al (20) for eutectic Sn-37Pb tested at 40°C. They found that dwell periods up to one hour, at one or both strain limits, had no influence on endurance compared to that measured for continuous cycling (Figure 5-22). As described previously, differential strain rate cycles were profoundly damaging to life. Such conflicting findings clearly demonstrate the need to inspect published results carefully.

The strain controlled fatigue behaviour of Sn-37Pb, Sn-3.5Ag and Sn-0.5Cu alloys has been examined at room temperature and 75°C (26). Figure 5-24 shows a composite Coffin-Manson plot for all data produced during continuous cycling. These alloys, in the as-cast condition, exhibit broadly similar fatigue characteristics. A temperature increase of 50°C produces a slight reduction in life, and the trends associated with strength and ductility described earlier are apparent. For example, the silver-containing alloy is inferior at short lives due to its lower ductility, but appears to exhibit the longest life (for $N_f > 5000$ cycles) due to its strength. The converse applies for the Sn-0.5Cu alloy. The incorporation of a dwell at any location in the strain cycle usually produces a reduction in fatigue life in Sn-37Pb, Sn-3.5Ag and Sn-0.5Cu (Figure 5-25). For dwells of 100s duration, the maximum life debit is about five times. The Sn-0.5Cu alloy is the most resilient and, exceptionally, for short dwells of 10s an enhancement of life relative to that observed for continuous cycling results. A key question, particularly for alloys in the as-cast condition, is the outcome of dwell periods of much longer duration when microstructural changes will occur during the dwell.

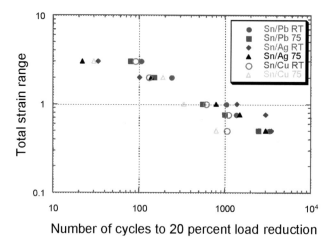

Figure 5-24. Coffin-Manson plot for lead-free alloys during continuous cycling (26)

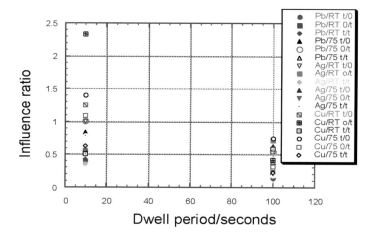

Figure 5-25. Effect of dwells of fatigue endurances of lead-free alloys (26)

4.4 Performance Under Fully Controlled TMF

The amount of data relating to solders is very limited, but knowledge of the shape of the hysteresis loop provides an insight into the key parameters in failure. In addition, examination of the behavioural trends in other engineering alloys during

TMF at similar homologous temperatures is indicative of possible responses in solders. In general, the stress range in TMF is greater than that for equivalent isothermal cycling (27) and the peak stresses are asymmetric about zero. Out-of-phase cycles are associated with a positive mean stress, and in-phase cycles result in a negative mean stress.

Steels and nickel alloys for use in power generation have been extensively examined during thermomechanical fatigue (28). Four characteristic patterns of behaviour between fatigue life, N_f, and the phase difference between temperature and strain have been observed.

1 $(N_f)_{IP} < (N_f)_{OP}$ at low strain ranges (low alloy steels - cast).

2 $(N_f)_{OP} < (N_f)_{IP}$ at low strain ranges (low alloy steels – forged).

3 $(N_f)_{OP} \approx (N_f)_{IP}$ for all strain ranges (stainless steel).

4 $(N_f)_{IP} < (N_f)_{OP}$ at high strain ranges (nickel superalloy).

The explanation for this wide, and sometimes apparently contradictory behaviour, is simply that the dominant mechanisms of failure may change. For example, fatigue or creep may dominate singly, or they may interact. Fracture may be transgranular or intergranular and so on. Each individual mechanism will have different sensitivities to the test parameters and the material microstructure.

The effect of hold periods at maximum temperature has been shown to have little effect on endurance during out-of-phase cycling for steels (28,29) but a substantial reduction results for in-phase fatigue. As more data on TMF of solders become available, it will be important to appreciate this and the need to acquire an understanding of the prevailing dominant failure mechanisms.

The available results for solder alloys are generally coincident with those above. The presence of an in-phase thermal cycle substantially reduces the endurance of a Pb-3.5Sn solder with respect to that observed isothermally at the maximum temperature of cycling (30).

Contrary to most findings for isothermal fatigue, a reduction in frequency resulted in an increase in TMF life. Similar observations, with regard to the comparison between TMF and isothermal fatigue life have been reported for Sn-37Pb, even though the tensile portion of the TMF hysteresis loop and that for the maximum temperature were similar (31).

4.5 Summary of Fatigue Behaviour

The eutectic Sn-37Pb and most lead-free alloys exhibit cyclic softening during high strain fatigue. When evaluated on the basis of a fixed definition of failure, the variation in fatigue endurance is not great, although the intrinsic strength and ductility effects appear at the extremes of the strain ranges. Conflicting evidence exists regarding the influence of hold periods at maximum strain. Some workers report substantial life debits, while others have found little effect. Differential strain rates, particularly slow-fast, can lead to life reductions in excess of an order of magnitude. Frequency has been reported to have the opposite effect on TMF than is usual during isothermal cycling. Until these apparent inconsistencies are understood, the fundamental 'physics of failure' approach to modelling and prediction is precarious.

5. CONCLUSIONS

A wide range of mechanical properties may affect joint performance in service. Even when assessed in the bulk form, the measured values are sensitive to a number of parameters, such as temperature, strain rate, time and prior history. The available data, particularly for lead-free solders, are far from complete and sometimes contradictory. There is a lack of understanding about the dominant failure mechanisms and how these might change. Substantial variations in any property may exist, which makes it imperative, in design and life prediction, to select values obtained under conditions appropriate to service. In the following chapter, the mechanical performance of joints, which have been additional complexities of constraint and the intermetallic layer, is considered.

6. REFERENCES

1   WJ Plumbridge, J Matls Sci., 1996, 31, 2501.
2   KP Jen and JN Majerus, J Engrg Matls. and Technol., 1991, 113, 475.
3   M Sakane, H Nose, M Kitano, H Takahashi, M Mukai and Y Tsukada, Int. Congress on Fracture, 10, Hawaii, December, 2001.
4   WK Jones, Y Liu, MA Zampino, G Gonzalez and M Shah, Design and Reliability of Solders and Solder Interconnections, (Ed R K Mahidhara et al), TMS, 1999, 85.
5   Solder Research Group (Update Pamphlet), 2001
      www.tech.open.ac.uk/materials/srg/srg-hp.html
6   WJ Plumbridge and CR Gagg, J. Electronic Matls., 1999, 10, 461.
7   WJ Plumbridge and CR Gagg, TMS Conference on Lead-bearing and Lead-free Solders, Columbus, Ohio, October 2002.
8   WJ Plumbridge and CR Gagg, British Association of Brazing and Soldering (BABS) Conference (Solihull, UK), 1996.
9   WJ Plumbridge, Int. Conf. On Fracture (ICF9), Adv. In Fracture Research, 1998, 167.
10   WJ Plumbridge, Engineering Against Fatigue, (Eds Benyon et al), Balkema, Rotterdam, 1999, 539.
11   WJ Plumbridge, CR Gagg and S Peters, J Electronic Matls, 2001, 30, 1178.
12   WJ Plumbridge and S Cooper Symposium on Pb-free and Pb-bearing Solders, TMS Fall Meeting, Columbus, Ohio, October, 2002.
13   FA Mohamed and TG Langdon, Philos, Mag., 1975, 32, 697.
14   WJ Plumbridge, Y Kariya and CR Gagg, Seventh Symp. On Microjoining and Assembly Technology in Microelectronics' (MATE 2001), Yokohama, February, 2001
15   Fatigue Database on Tensile and Low Cycle Fatigue Properties of Sn-37Pb and Sn-3.5Ag Solders, Society Materials Science, Japan (Kyoto), 2001.
16   WJ Plumbridge, Materials at High Temperatures, 2000, 17, 381.
17   CH Raeder, RW Messler Jr and LF Coffin, J Electronic Matls, 1999, 28, 1045.
18   R Sandström, J-O Osterberg and M Nylon, Matls. Sci. and Techn., 1993, 9, 811.
19   XQ Shi, HLJ Pang, W Zhou and ZP Wang, Int. J. Fatigue, 2000, 22, 217.
20   H Nose, M Sakane, H Tsukada and H Nishimuta, J Electronic Packaging, 2003, 125, 59.
21   T Morihata, Y Kariya, E Hazawa and M Otkuka, Seventh Symp. Microjoining and Assembly Technology in Electronics (MATE 2001), Yokohama, February, 2001.
22   Y Kariya and M Otksuka, J Electronics Matls, 1998, 27, 866.
23   Y Kariya and M Otsuka, J Electronics Matls, 1998, 27, 1229.
24   S Vaynman and ME Fine, Proc. 2nd ASM Inst. Electronic Materials and Processing Conf., 1989, (Ed W T Shieh), 255.
25   HD Solomon, 38th Electronic Components Conf., IEEE, 1988, 7.
26   JE Moffatt and WJ Plumbridge, Tenth Int. Congress on Fracture (ICF 10), Hawaii, December 2001.

27 K Kuwabara, A Nitta and T Kitamura, ASME, Int.Conf. on Advances in Life Prediction Methods, 1983, 131.
28 K.Kuwabara and A Nitta, CRIEPI Report, E277005, 1977.
29 A Nitta and K Kuwabara, High Temperature Creep-Fatigue, (Elsevier Applied Science), 1988, 203.
30 LR Lawson, ME Fine and DA Jeannotte, Metall. Trans, 1991, 22A, 1059.
31 EC Cutiongco and DA Jeannotte, Proc. M E Fine Symp., 1990, Detroit.

CHAPTER 6

MECHANICAL BEHAVIOUR OF SOLDER JOINTS

1. INTRODUCTION

Having examined the mechanical properties of bulk solders, attention is now turned to the behaviour of solder joints, which may be regarded as composite structures, with the solder as the common ingredient. The interfaces of the solder with the substrate consist of intermetallic compounds, such as Cu_6-Sn_5 and Cu_3-Sn, and within the body of the solder, particles of these IMCs may exist after fragmentation from the main layer. So, even the nature of the solder differs from its bulk form. Apart from the challenges of variability and uncertainty discussed in the previous chapter, specification of *joint* properties is further complicated by the existence of a near-infinite number of permutations of components, materials, geometry and processing history. Data obtained are highly specific. With this *caveat*, the present chapter considers a representative selection of the information that is available. Strength, as exhibited during monotonic, cyclic and sustained loading, is initially examined, followed by TMF behaviour. An underlying question is what, if any, is the correlation between bulk and joint behaviour?

2. STRENGTH OF JOINTS

2.1. Monotonic

The majority of studies on model joints (1-6) have utilised the pin-in-ring configuration due to its close proximity to pure shear and its ease of manufacture. In addition, the scope for limiting its dimensions enables rapid cooling rates and realistic microstructures to be achieved. There is usually less scatter in the measurements than observed in double-lap joints (2). However, scatter in reported data for strength may be as large as a factor of two.

The trends in the monotonic behaviour of joints mirror those exhibited by bulk solders. Figure 6-1 reviews data for the three alloys about which most information is available, and while direct comparison is not viable, the trends observed for bulk samples persist. Sn-0.7Cu joints have shear strengths similar or lower than joints manufactured with Sn-40Pb, and the silver-containing alloy joint possesses the highest shear strength.

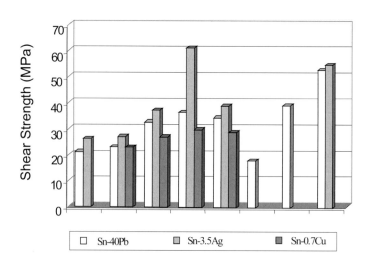

Figure 6-1 Review of data for the shear testing of Sn-40Pb, Sn-3.5Ag and Sn-0.7Cu model solder joints at room temperature.

Strength diminishes with slower straining rates and increasing temperatures. Increasing the strain rate by a thousand times typically doubles the shear strength of both lead-containing and lead-free pin-in-ring joints at room temperature (5). However, for high lead (lead content above 90 per cent) solders, this effect is reduced (5). Similar consequences arise when the test temperature is reduced from 100°C to ambient, with the lead-rich alloys again being less sensitive (Figure 6-2).

Figure 6-3 shows the variation of shear strength of Sn-40Pb and Sn-3.5Ag joints with temperature and strain rate (as denoted by crosshead speed). Note the reversal of ranking under the higher strain rate/high temperature conditions.

The specific joint geometry has an effect on measured strength. Shear strengths for the 'butt' joint configuration (asymmetric four-point bend) (7) are consistently higher than those for the pin-in-ring joints (8), particularly for the intrinsically stronger silver-containing alloy (Figure 6-4).

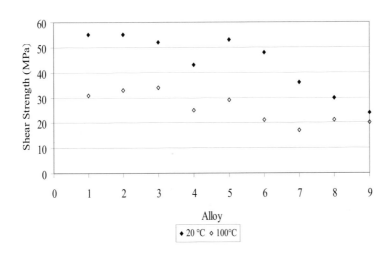

Figure 6-2 Effect of temperature on shear strength(5)
Alloy identification: (1) Sn-3.5Ag; (2) Sn-36Pb-2Ag; (3) Sn-40Pb; (4) Sn-58Pb-2Sb; (5) Sn-5Sb;
(6) Sn-60Pb; (7) Sn-90Pb; (8) Sn-93.5Pb-1.5Ag; (9) Sn-97.5Pb-1.5Ag
Deformation rate 50mm/min.

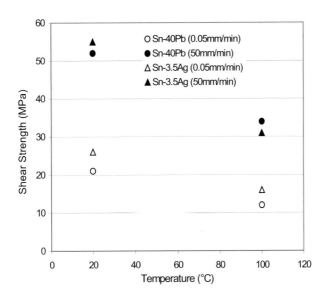

Figure 6-3 Variation of shear strength of Sn-40Pb and Sn-3.5Ag 'pin in ring' joints with
temperature at two deformation rates (5).

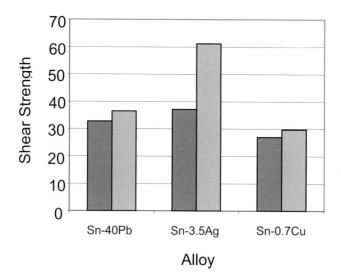

Figure 6-4 Variation of shear strength with joint geometry (7,8).
(Dark; pin – in-ring; Light; Butt AFPB)

The effect of solder thickness differs according to the value of the thickness itself and that of the IMC layer. Little difference is observed in the strength of joints having solder thicknesses ranging between 0.03 and 0.26 mm (2). However, with a solder layer thickness 0.7mm, measurements of the shear strength of Sn-40Pb pin-in-ring joints indicate that there is an optimum thickness of the IMC layer, typically about 1.5μm (1,3, 9). Above this, the strength deteriorates (Figure 6-5).

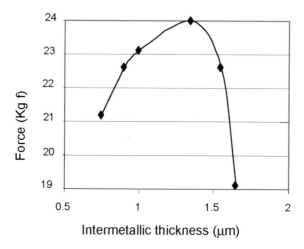

Figure 6-5 Effect of intermetallic thickness on shear strength of soldered joints (3).

2.2 Cyclic

As indicated previously, high strain fatigue is not particularly sensitive to microstructure or strength, since it is controlled primarily by ductility. In contrast, under low stress/high cycle conditions, such as vibration, both microstructure and strength are expected to have a strong influence on performance.

Using a straddle shear fatigue system, Kariya et al (10) have recently evaluated the influence of silver content on the isothermal mechanical fatigue endurance of Sn-Ag-Cu flip chip interconnects. Similar to their studies on bulk alloys (11), the effect of silver could be explained in terms of its influence on strength and ductility. The higher silver contents (3~4 per cent) produced optimum strength and high cycle fatigue resistance whereas the weaker, but more ductile, one percent silver alloy gave maximum high strain fatigue endurance (Figure 6-6). For any strain range, the spread of fatigue lives falls within a factor of five – so the effect is fairly small.

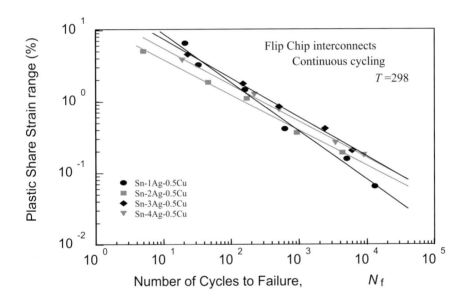

Figure 6-6 Relationship between fatigue life and shear plastic strain range for flip chip joints (10).

Isothermal mechanical fatigue of 0.5 and 0.8 mm pitch CSP and a PCB indicates little difference between the performance of Sn-3.5 Ag-0.75 Cu and Sn-37Pb joints (12). However, when subjected to prior ageing (20 days at 120°C) the lead-bearing alloy was considerably inferior. The finer pitch also exhibited the shorter life.

In model butt joints containing a centrally notched solder (Sn-37Pb) layer, fatigue crack growth under shear (Mode II) loading may be correlated with the stress intensity range, ΔK, the ΔJ integral or the C^* integral – as is the case for tensile Mode I loading. However, the power term for ΔK is 2 to 3 times higher (13). The crack path exhibited no clear microstructural preference although it grew in the direction of the solder interface.

Subsequent propagation was in the solder adjacent to this layer. The nature of the base material can also be influential. Fatigue crack growth in a Sn-3.5Ag solder deposited on copper is more rapid than when the same alloy is deposited on electroless nickel (14). This is attributed to coarse Ag_3-Sn particles present in the former case. Chapter 13 considers crack growth in further detail, in the context of life prediction.

2.3 Creep

Most actual and model joints have been creep tested under shear conditions whereas the vast majority of data on bulk samples has been obtained from tensile loading. While the classic relationships from equations 4.1 and 4.2

$$\tau = \frac{\sigma}{2(1+v)} \text{ and } G = \frac{E}{2(1+v)} \tag{6.1}$$

are commonly used to convert the measurements for comparative purposes, there is no evidence to substantiate their use for solder alloys at low strain rates. In lap or pin-in-ring joints, the shear stress, τ, is not distributed uniformly in the shear direction, and an average value involving the minimum loading area is utilised.

With the enormous number of possible parameters that may be examined, the available information at present can best be described as fragmentary. In an extensive study, Clech (15) has collated and analysed data on lead-free solders in bulk and joint form. Employing a sophisticated curve-fitting programme for the regression analysis, 'master curves' have been produced for particular alloys that indicate a band of behaviour and identify possible 'rogue' data. Figure 6-7a and b show these for Sn-3.5Ag alloys. Data from several sources and a wide range of conditions (Figure 6-7a) are consolidated into a band around (in this case) a hyperbolic sine function. (Figure 6-7b), ie

$$\dot{\varepsilon}_m = A[\sinh(B.\sigma)]^n \exp\frac{-Q}{RT} \tag{6.2}$$

The centre line in Figure 6-7b is a plot of $\dot{\varepsilon}\exp\left(\dfrac{Q}{RT}\right)$ v σ, and significant scatter is apparent.

Within the limitations of the available data, this approach can be applied to joints. Figures 6.8a and b show the analysis of the data for Sn-3.5Ag lap and pin-in-ring joints. The scatter is usually greater than that for the bulk material, and the form of the equation is similar apart from the replacement of tensile by shear stress terms.

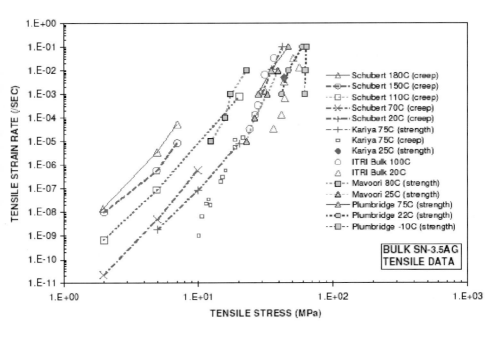

Figure 6-7a. Isothermal Tensile Creep Data for Bulk Sn-3.5Ag solder (15)

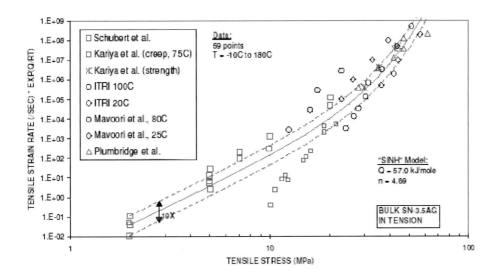

Figure 6-7b. Curve-fitting of bulk Sn-3.5Ag tensile creep data to hyperbolic sine mode (15).

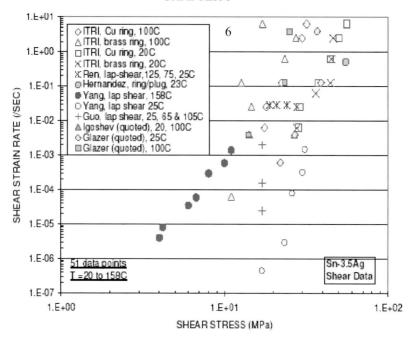

Figure 6-8a. Creep data for the Sn-3.5 Ag lap shear and pin-in-ring joints(15)

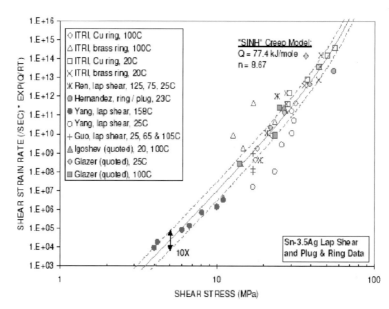

Figure 6-8b. Fit of "sinh" model to Sn-3.5Ag lap shear and pin-in-ring data (15)

Trends that are not apparent from individual data sets can also be predicted from the consolidated master curves. For example, inspection of the master curves for Sn-37Pb, Sn-3.5 Ag and Sn-Ag-Cu at various temperatures suggests a change in the relative creep resistance of these alloys. At high stresses (above - 40MPa for Sn-3.5 Ag) minimum strain rates become higher in the lead-free alloys than in Sn-37Pb. For temperatures of 100, 25 and -55°C, the transition strain rates were 5 x 10^{-1}, 1 x 10^{-3} and 5 x 10^{-7} s^{-1} respectively. Such high stress levels could be encountered in accelerated testing, and it is pertinent to note that the lives of lead-free chip resistor assemblies during thermal cycling have been found to be lower than those in Sn - 37Pb assemblies (16).

Other indicated trends include: a higher stress exponent in the stress-strain rate relationship for joints relative to bulk alloys; a correlation between creep in ceramic chip carriers and in lap or pin-in-ring joints; and an equivalence between the shear behaviour of flip chip and ball grid array configurations, although the former appear less creep resistant than ceramic chip carriers at room temperature. Silver-containing alloys in any form are more creep resistant in compression than in tension, but around the eutectic composition, creep performance is relatively insensitive to silver content (15).

At high stress levels, equivalent to rupture times of a few hundred hours, the creep behaviour of pin-in-ring model joint specimens of Sn-3.8Ag-0.5Cu has been shown to be similar to that of bulk specimens at room temperature and 75^{0}C, when assessed in terms of time to rupture (17). At lifetimes in excess of this, the joint sample was inferior, and this was attributed to the greater instability of is microstructure. Significant coarsening of the IMC particles within the body of the solder was observed, together with a transition in the Norton exponent from above ten to three. A similar mechanism change was not apparent in the bulk specimen data under the conditions examined (lifetimes up to 1000 h), and the stress exponents in Norton's law varied between 18 and 10 as the test temperature changed from -10 to 125ºC.

It cannot be emphasised strongly enough that these observations should not be regarded as universally applicable trends. They do, however, point to characteristics which require confirmation when a specific joint configuration and its operating conditions are under consideration. The pitfalls associated with intuitive extrapolation from lead-tin alloys to lead-free systems are well illustrated here. It has been observed that cooling rate has the opposite effect on the tensile and creep behaviour of joints. (18,19). To understand this, the substantial metallurgical differences between lead-containing and lead-free solder alloys must be appreciated. Apart from the large difference in solute content, the microstructure of Sn-37 Pb is essentially a tin matrix containing softer lead-rich phases, whereas in Sn-Ag, for example, the tin matrix is reinforced by small stiffer particles of Ag$_3$Sn. It is hardly surprising that different mechanical responses are exhibited. In the Sn-37 Pb alloy it is suggested that grain boundary sliding is the dominant creep deformation mechanism (15).

2.4 The Scatter Problem

The numerous variables associated with nominally identical tests and the absence of appropriate standards have been mentioned previously. It has been acknowledged that although the situation regarding determination of bulk properties is difficult, that for joints is substantially worse. Any trend in one data set is likely to be challenged or rebutted in results from other sources. All information should be viewed with this in mind.

Scatter in data may also arise from other joint-specific features, such as shape, size, volume or extent of dissolved termination metals. A flip chip joint may be up to 200 times smaller than a BGA which, in turn, may be as much as a 1000 times smaller than a bulk sample.

Microstructure developed during processing can have a substantial affect on performance and is determined largely by the cooling rate. Recommended maximum cooling rates in service are 6 K s^{-1}, which is much slower than often reported values for as-cast bulk laboratory specimens ($\sim100 \text{ K s}^{-1}$). It is inevitable that microstructures in these cases will be different and the rigour of the comparison will be compromised. Along similar lines, actual joints do not experience a "stabilisation anneal" (e.g. 1h at 100°C) which is utilised in many laboratories prior to evaluation of mechanical behaviour. The potential significance of this was reported in Chapter 5.

The joint volume and shape, and the nature of the neighbouring materials also affect the cooling rate experienced by the solder alloy. Dramatic differences in microstructure may arise and these, in turn, can affect mechanical performance. For example, in the Sn-3.5Ag system, water quenching (24 K s^{-1}) results in a microstructure of tin-rich dendrites surrounded by a fine eutectic mixture of spherical Ag_3-Sn in a tin-rich matrix (20). Air cooling (0.5 K s^{-1}) produces a coarser arrangement of the same constituents, whereas furnace cooling (0.08 K s^{-1}) results in a eutectic mixture of Ag_3-Sn needles in a tin-rich matrix. Such variations have little effect on Young's modulus provided that the IMC particles are randomly aligned. However, an increased cooling rate causes an increase in yield strength and work hardening rate, at the expense of ductility. The tensile strength is unchanged. (This demonstrates the benefits of measuring several mechanical properties, since the constitutive equations would differ although the single measurement would indicate no change.) Similarly, an increased cooling rate improved creep resistance, giving a higher 'n' value in the strain rate-stress expression. This was attributed to the greater resistance to dislocation motion provided by the finer microstructure and the Ag_3-Sn particles. Under the conditions examined, grain boundary sliding played little part in the creep of this alloy.

All other factors being the same, the correlation between bulk and joint strength depends upon the relative strength of the solder and the IMC layer, and the thickness of the latter (21). In large soldered joints, in which the thickness of the IMC is low compared with that of the solder layer, there is little difference because crack growth occurs within the solder itself (21). Kitano et al (22) confirm this with their findings on pin-in-ring joints with a 5mm solder layer subjected to thermal cycles. However, as miniaturisation continues, intermetallics will play an increasingly important role in reliability.

3. THERMOMECHANICAL FATIGUE IN JOINTS

Having introduced the fundamentals of thermomechanical fatigue in earlier chapters, its occurrence in model and actual joints is now considered. The advantages and shortcomings of each joint configuration were considered in Chapter 4, and should be recalled here.

3.1 Model Joints

Pao et al (24) have used model joints, comprising alumina and a 2024-T4, aluminium alloy, to evaluate the thermal cycling performances of a range of solder alloys between 40 and 140°C. They assert that due to bending, such a configuration is neither under stress control nor strain control, and that both stress relaxation and creep occur during dwell periods. Under the test conditions above, the life of the joint with the Sn-37Pb alloy was between 4 and 10 times greater than that with high -lead alloy solder or lead-free solders, such as Sn-3Cu, Sn-4.0Cu-0.5Ag. None of the currently more popular lead-free compositions was examined.

Employing a three-bar system of stainless steel and Invar, cycled between 30 and 125°C, Liu and Plumbridge (25) have demonstrated that the Coffin-Manson and energy-based expressions correlate with endurance data for a eutectic Sn-Pb solder. Failure occurs by crack growth in the solder adjacent to the IMC layer. Using the ternary, Sn-Ag-Cu alloy, cracks grow closer to this layer and often at the interface (26).

A test system capable of cycling model butt joints in shear has been employed to monitor the development of stress and strain during the actual formation of the joint. (27, 28). With a Sn-42Bi solder alloy, significant stress (~30 MPa) and shear strain (~30 per cent) were generated during cooling from the solidification temperature (138°C) to 0°C. The maximum stress level after the high temperature stage of the cycle was about 5MPa. Subsequent thermal cycling between 0 and 100°C produced a reduction in stress range and a gradual increase in shear strain range until the fully constrained limit of 38.5 per cent was attained. To accommodate this variation in shear strain range throughout life, an effective strain range, $\Delta\gamma_{eff}$, (the average of the stable strain range and the strain range at failure) gave a good fit in Coffin-Manson plots.

3.2 Actual Joints

In this context, joints may be regarded in two categories; leaded and leadless, i.e. with, and without leads. Although the leadless chip carrier offers many benefits, it does depend totally upon a column of solder to perform satisfactorily over a wide range of conditions. Analytically, the strains developed in this configuration are similar to those derived previously in Chapter 3. The introduction of a compliant lead between the chip carrier and the substrate significantly reduces the thermal mismatch problem between the package and board, and the plastic strain range experienced by the solder (29). However, the elasticity of the leads may prolong the duration of creep stresses in the solder and induce elements of creep strain which are additional to the cyclic mismatch plasticity. In effect, the benefits of lead flexibility may be nullified. The stiffness of the leads is crucial. Englemair (30) has developed an expression for the strain range, $\Delta\gamma$, in terms of lead stiffness, K:

$$\Delta\gamma = C\left(\frac{K\Delta D^2}{Ah}\right) \tag{6.3}$$

where C is a constant, ΔD the lead displacement range, A the effective joint loading area, and h the effective joint height. If the leads are flexible (low K values) stress relaxation is slow, which allows prolonged periods for the creep strain to accumulate.

During mechanical cycling, the frequency is also significant; as it falls so does the strain range (Fig 6.9). It can be shown that the strain range is given by:

$$\Delta\gamma = C\frac{K^n\Delta D^n}{A^ng^p}t_d \tag{6.4}$$

where the stress in the solder, τ, is $K\Delta D/A$, n is the creep exponent of the solder, g the grain size, p the grain size exponent, and t_d is the half cycle dwell time. For flexible leads, the strain range is substantially controlled by the constitutive properties of the solder.

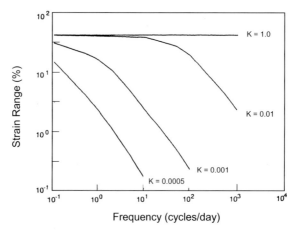

Figure 6-9 Influence of frequency and relative spring stiffness, K, on the strain range in the solder
(29)

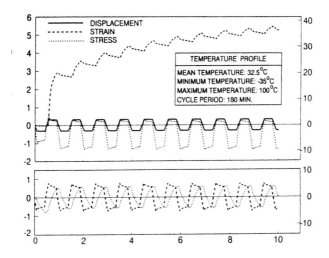

Figure 6-10 Shear stress and strain response to TMF (upper) and isothermal fatigue (lower)(29)

During thermomechanical fatigue, creep ratcheting can occur, due to the strong temperature dependence of creep during different phases of the loading cycle. More rapid and extensive creep occurs on the higher temperature side of the cycle, whereas on the low temperature side the solder is more creep resistant and lead-deflection accommodates most of the strain. Chapter 5 provides quantitative information on this. On reheating, the force from the deflected lead produces considerably more strain due to the higher temperature. Eventually, equilibrium is reached which is determined by the stiffness of the lead and the solder properties (Fig 6.10). A general guideline is that the lead stiffness should be less than 0.1 per cent of that of the solder. While creep ratcheting is usually self-arresting, it may produce damage which accentuates subsequent fatigue processes. For example, it may cause rapid initiation of cracks. So while leaded components experience smaller cyclic strain ranges, a compromise is involved in the selection of lead stiffness in order to avoid excessive generation of strain from creep ratcheting.

On any printed circuit board, there is generally a wide range of components and interconnections. Profiles of individual joints differ as do their associated stress states. Direct measurement of stress and strain is difficult. A common means of evaluating performance is to examine the system after interrupted thermal cycling and gradually build up a picture of damage accumulation. Using this approach, it has been demonstrated that for a plastic quad flat pack (PQFP), with 256 leads and Sn-37Pb solder, cracks initiate at the heel of the solder fillet during thermal cycling (31). Further initiation occurs in the toe region and finally, cracks extend through the solder between the copper lead and the board.

Measurements of deflections in the PWB and a ceramic chip carrier have been used to infer stress-strain conditions within a joint (32), and these have been confirmed by computer simulation (33).

Perhaps the most influential study in recent years has been the IDEALS (*Improved Design Life and Environmentally Aware Manufacturing of Electronic Assemblies by Lead-Free Soldering*) project which concluded that tin was probably the only possible base for new solder alloys (34). Test circuits containing resistors and capacitors soldered with eutectics based on Sn-3.8 Ag, Sn-0.7Cu and Sn-3.8Ag-0.7Cu (and minor modifications of them) were subjected a series of thermal cycling tests with Sn-40Pb and Sn-62Pb as comparator alloys. The thermal tests included shock, power cycling and chamber cycling. The principal conclusions were that there is no significant difference in terms of performance under fluctuating temperatures between the lead-containing and the lead-free alloys, and that the reflow profile was not significant. Figures 6.11 and 6.12 demonstrate the broad similarities in performance for the thermal and power cycling. The ternary Sn-3.8Ag-0.7Cu was identified as the most likely universal replacement alloy.

Kariya and co-workers (35, 36) have demonstrated the effect of copper and bismuth additions to a Sn-3.5Ag solder joint on a quad flat pack assembly. In terms of the lead pull-out strength, 0.5 and 1.0 per cent copper resulted in an improvement over those in the binary alloy (Fig 6.13) which gradually reduced with thermal cycling between 0 and 100°C. While bismuth additions up to 7.5 per cent were initially beneficial, subsequent thermal cycling caused a pronounced fall in pull-out strength to less than a half of that in the binary alloy and its value prior to thermal cycling. Pull-out strengths in the binary Sn-3.5Ag were unaffected by up to 1200 cycles, while those in the lead-tin eutectic were reduced by some 30 per cent by the same number of cycles.

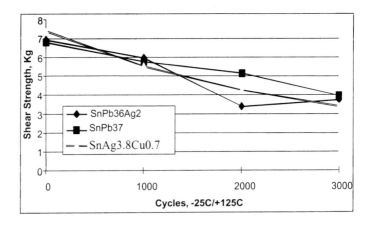

Figure 6-11 Effect of thermal cycling on joint strength (34)

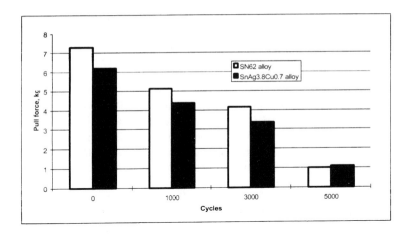

Figure 6-12 Effect of power cycling on joint strength (34)

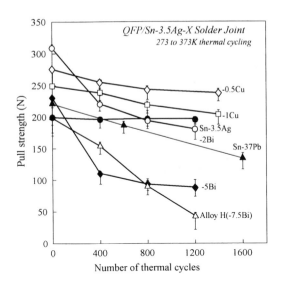

Figure 6-13 Effect of bismuth additions on thermal cycling performance (35)

The thickness of the IMC layer has been shown to have an effect on the thermal fatigue life of a Sn-37Pb SMT joint (37). During cycling between -35 and 125°C, the endurance was roughly halved as the IMC layer thickness increased from 0.8 to 1.4 μm. At greater thicknesses, up to 2.8 μm, fatigue life fell only slightly. A transition in fracture path from the solder to the interface with the IMC occurred as the latter grew.

4. CONCLUSIONS

This chapter has illustrated the importance of mechanically testing joints that are complex in comparison to bulk solder materials and exist in many forms. To correlate joint and bulk behaviour would be a valuable contribution to design and life prediction. The evidence to date indicates that trends exhibited in bulk materials persist to joint behaviour, although direct correlation seems remote until a much better understanding has been achieved.

It cannot be emphasised too strongly that all data are limited to the conditions under which they were derived. Both in terms of relevance to actual performance under service and comparability (or ranking) of alloys, extrapolation of the findings to a broader spectrum is unsubstantiated.

While joints with leads are beneficial in terms of the reduced strain range experienced by the solder, lead flexibility and subsequent stress relaxation should be considered to avoid elimination of this advantage.

The diversity of methods for assessing the thermomechanical fatigue behaviour exacerbates the problem of comparability of results between laboratories. Experiments involving complete boards provide valuable empirical information regarding the effect of specific parameters, but their generic contribution is limited. The IDEALS project has identified the ternary Sn-3.8Ag-0.7Cu as the most likely universal replacement alloy for Sn-37Pb, and this has been confirmed in a SMART Group poll of industry in the UK (38), when 86.5 per cent of the respondents considered this alloy to be the most popular.

5. REFERENCES

1   SF Dirnfield and JJ Ramon, Supplement to The Welding Journal, 1990, (Oct).
2   CJ Thwaites and R Duckett, Review de la Soudre, 1976 ,4, 1.
3   JJ Ramon and SF Dirnfield, Welding Journal, 1988, 67,10.
4   JC Foley, A Glickler, FH Leprovost and D Brow, J Electronic Matls, 2000, 29, 1258.
5   KR Stone, R Duckett, S Muckett and ME Warwick, Mechanical Properties of Solders and Solder Joints, ITRI, Uxbridge,1983.
6   F Gillot, Fatigue Behaviour in Electronic Solder Joints, Swedish Inst. For Metals Research, (Stockholm), 1997, 46.
7   IE Anderson, TE Bloomer, RL Tepstra, JC Foley, BA Cook and J Harringa, Int. Summit on Lead-free Assemblies, Minneapolis, MN, 1999.
8   GSA Shawki and AAS El-Shabbagh, Welding Research Suppl., 1975.
9   PG Haris and KS Chagger, Soldering and Surface Mount Techn., 1998, 10,38.
10 Y Kariya,T Hosei, S Terashima, M Tanaka and M Otsuka, J Electronic Matls, to be published.
11 Y Kariya and M Otsuka, J Electronic Materials, 1998, 27, 1229.
12 Y Kariya, T Hosoi, Y Tanaka and M Otsuka, Eighth Symp. on Microjoining and Assembly Technol. in Electronics (MATE 2002), 2002, 8, 431.
13 Z Guo and H Conrad, J Electronic Packaging, 1993, 115. 159.
14 P Liu and J K Shang, J Electonic Matls, 2000, 29,622.
15 JP Clech, 'Review and Analysis of Lead-free Solder Material Properties', in NEMI Lead-free Project Book, Chap 13, IEEE/Wiley, 2003.
16 G Swan, A Woosley, N Vo and KT Koschmieder, Proc. IPC SMEMA Council APEX 2001, Paper LF"-6.

17 S Cooper and WJ Plumbridge, TMS Annual Conf. on Lead-bearing and Lead-free Solders, Columbus (Ohio), October, 20002.
18 Z Mei, JW Morris, MC Shine and TSE Summers, J Electronic Materials, 1991,20, 599.
19 Z Guo and H Conrad, J Electronic Packaging, 1996, 118, 49.
20 F Ochoa, JJ Williams and N Chawla, J of Materials, 2003, 55, 56.
21 JW Morris, D Tribula, TS Summers and D Grivas, Solder Joint Reliability, (Ed J H Lau) Van Nostrand and Reinhold, 1991,225.
22 M Kitano, T Kumazawa and S Kawai, Adv. In Electronic Packaging, ASME, 1992,1, 308.
23 TA Woodrow, IPC and JEDEC Int. Conf. On Lead-Free:Electronic Components and Assemblies, San Jose, May, 2002.
24 YH Pao, S Badgley, R Govila and E Jih, Fatigue of Electronic Matls., ASTM STP 1153, (Eds SA Schroeder and MR Mitchell), ASTM, Philadelphia, 1994, 60.
25 XW Liu and W J Plumbridge, Matls Sci and Engrg, 2003 to be published.
26 XW Liu and W J Plumbridge to be published.
27 PE Hacke, AF Sprecher and H Conrad, J Electronic Packaging, 1993.15, 153.
28 CH Raeder, RW Messler and LF Coffin, J Electronic Materials, 1999, 28, 1045.
29 RG Ross, Jr and L-C Wen, Thermal Stress and Strain in Microelectronics, (Van Nostrand-Reinhold), New York, 1993, (Ed J H Lau), 607/
30 W Englemair, Solder Joint Reliabiltiy - Theory and Application, (Van Nostrand, New York) 1991, 543.
31 CK Yeo, S Mhaisalkar, HLJ Pang and A Yeo, J Electronics Manuf., 1996, 6, 67.
32 PM Hall, IEEE Trans Components, Hybrids Manuf. Technol., 1984, 7, 314.
33 RG Ross, JEP, 1994, 116, 69.
34 MR Harrison, JH Vincent and HAH Steen, Soldering and Surface Mount Technol., 2001, 13 21.
35 Y Kariya, Y Hirata and M Otsuka, J Electronic Matls, 1999, 28, 1261.
36 K Warashina, Y Kariya, Y Hirata and M Otsuka, Sixth Symp. on Microjoining and Assembly Technology in Electronics' (MATE 2000), Yokohama, February, 2000.
37 PL Tu, YC Chan and JK Lai, IEEE Trans Components, Packaging and Manufacturing Technology, Part B, 1997, 20,1.
38 SMART Group Survey, www.smartgroup.org , April 2001.

PART 2

THE MANUFACTURER'S PERSPECTIVE

The focus of this Part is entirely different from that of Part 1. It addresses the challenge produced by the forthcoming implementation of lead-free technology from the viewpoint of the manufacturer of electronic equipment. To set the scene and to inform those not specialising in electronics, the initial chapter describes common types of electronics components, their function and the manufacturer's requirements for efficiency and reliability.

Subsequent chapters explore the origins of potential defects that may arise during manufacture and assembly, and more importantly, how these features might be eliminated or minimised. Associated with this section, the techniques of failure analysis and the various items of equipment utilised in the diagnosis of failure modes, are considered.

Finally, Part 2 addresses the question that is currently foremost in the minds of industry - How can the transition to lead-free technology be made as efficiently and economically as possible, with no reduction in joint reliability? In this context, it is able to draw upon many of the factors that have appeared elsewhere throughout the Book.

CHAPTER 7

ELECTRONIC COMPONENTS AND THEIR CONSTRUCTION

1. THE COMPONENT PACKAGING REVOLUTION

During the 1960s, electronic products were large, heavy and power hungry. Early computers filled large rooms and were very slow. Equipment size was linked to the thermionic valve devices, which were predominately single function with limited integration of not more than one function per component package. The impressive improvements in materials and construction concepts of electronic equipment and components since then, has resulted in a minor industrial miracle. The computing power of a simple mobile cell phone, with the integration of millions of transistors onto a single piece of silicon, is already taken very much for granted.

This chapter considers the construction methods used to manufacture electronic devices. The information is not intended to give a functional explanation of each component type, but to raise the awareness of the construction and parametric details that are important during failure analysis. Today, component packaging technology is predominately driven by product down sizing and surface mount technology (SMT). SMT, in turn, is feeding the high-speed automated PCB assembly processes (Chapter 8).

There are a large variety of package types, from high pin count integrated circuit Ball Grid Arrays (BGAs) through to miniature two terminal passive devices. Mechanical construction is often forgotten in favour of the operating complexity and highly published microscopic diffusion geometries of the silicon. The challenge in constructing high pin count BGAs and other high density low profile packages, involves routing low profile fine pitch bond wires that are thinner than a human hair, at high speed production through puts, between the die and connecting pads of the external lead-frame connections. This achievement is matched by the construction of the less complex passive device, which is manufactured in extremely high volumes, exhibits high production yields and reliability, and sells to the end user at almost zero cost.

Component users are often responsible for introducing device failures, and failures attributed to the component manufacturer are rare. This is a tribute to modern automated construction methods, quality control, reliability testing and improvements in materials and final test methodology.

Surface mount components now dominate the printed circuit board assembly industry, constituting 75 percent of production during 2002. However, leaded parts will continue to dominate some areas, particularly where high power is required or surface mount technology is unable to provide device functionality. This chapter addresses both types of device but with the bias towards surface mount technology.

2. RESISTORS

Resistors and capacitors have undergone massive changes over the past few years, exhibiting vastly improved characteristics and ever decreasing dimensions. To meet the demands of miniaturisation, surface mount chip resistors and capacitors dominate the passive market, by volume. Figure 7-1 is a cross section view of a chip resistor, which is representative of all manufacturers construction methods. Construction is similar to the silicon wafer, with several hundred resistors assembled onto one square ceramic substrate. For example, during the initial manufacturing stage of 0805 chip resistors, 1360 resistor elements and inner terminations are printed onto a single piece of substrate six centimetres square, which is subsequently scribed and separated during the production process to single chip resistor elements.

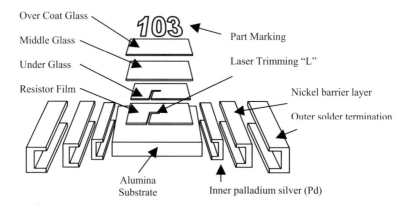

Figure 7-1. Construction details of a standard surface mount chip resistor.

Figure 7-2 illustrates the first step in the manufacturing process, with the screen-printing of the front and back inner silver palladium (Pd-Ag) terminations onto the substrate, followed by the printing of the thick film resistor base material. Screen-printing utilises a metal foil with a precision aperture cut into it to coincide with the areas where materials are to be deposited onto the substrate. For chip resistors, there are two screens, one for the inner termination and one for the resistor element. The screen is lowered and precisely located onto the substrate surface and the material in paste form is pressed through the apertures by a squeegee action. To form the resistor material, powdered ruthenium and glass are mixed with a binder to produce a paste of a known square ohmic value. The resistance value after firing will be 20 percent lower than the target value. Resistance is determined by the ruthenium content. The glass/ruthenium mixture is again screen-printed on top of the previously prepared inner silver terminations, and oven fired to form the base resistor element.

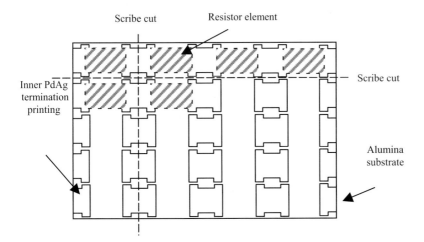

Figure 7-2. Chip resistor alumina base with inner termination and resistor element printing.

Each chip resistor is laser trimmed to the final value. Laser trimming is effective, fast and accurate with average trim time only a few milliseconds per resistor, as each strip is simultaneously trimmed on the substrate before separation.

After laser trimming, a further two layers of glass protection are printed and fired over the resistor element to prevent moisture ingression and corrosion. Part marking is then performed using a three or four digit code according to whether the tolerance is 5% or 1% respectively. Three digit part marking for a 15,000 ohm resistor (15K) would be 153 i.e. 15 and (3) three zeros. Four-digit marking for a 3,830 (3.83K) will be 3831 i.e. 383 and (1) one zero. Resistor tolerances down to 0.5% with temperature coefficient of resistance (TCR) +/-50 ppm/degree centigrade can be achieved. For higher precision, lower noise and improved characteristics, the circuit designer should chose metal or thin film type chip resistors which exhibit characteristics 0.25% tolerance and TCR <0.25ppm/degree C.

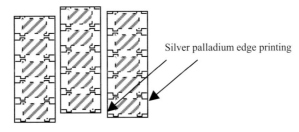

Figure 7-3. Resistor scribing and bar breakout to enable inner edge termination printing

At this stage, the ceramic substrate is scribed and broken into strips or bars where the resistor edge termination (silver palladium) is printed and fired Figure 7-3. There is no termination metal on the side edges of chip resistors, which prevents solder bridging between adjacent resistors during the soldering process. Each individual resistor is then scribed and separated from the bar. Finally, a nickel barrier layer is electroplated onto the Ag-Pd connection. This layer prevents migration of the silver inner termination into the outer solder during reflow soldering. After a washing cycle, the process is repeated to electroplate the tin-lead terminations (Sn-10Pb) onto the nickel to form the outer termination. To comply with the implementation of lead-free solders, this will be replaced with tin electroplated on to the nickel barrier layer.

Silver-palladium is used for the inner termination to produce a very low electrical resistance connection between the resistor ruthenium oxide element and the outer terminations. Surface mount chip resistors, although having specific manufacturers part numbers, are often described by their dimensions in hundredths of an inch or metric equivalents, e.g. a 1206 is 12 hundredths x 6 hundredths of an inch or (3216) in metric 3.2 x 1.6 mm. Downward trends are currently demanding 0201 resistors (0603) 0.6 x 0.3 mm for high-density application. Table 7.1, gives imperial, metric and true dimensions of currently available chip resistor series and future trends. A 0.5 x 0.25 mm chip resistor is already planned.

Table 7.1.Chip resistor body size

Imperial (hundreds of an inch)	Metric (mm)	Actual dimensions (mm)
2512	6432	6.3 x 3.2
2010	5025	5.0 x 2.5
1210	3225	3.2 x 2.5
1206	3216	3.2 x 1.6
0805	2012	2.0 x 1.25
0603*	1608	1.6 x 0.8
0402	1005	1.0 x 0.5
0201	0603*	0.6 x 0.3
0101	05025	0.5 x 0.25

Beware of confusion here between metric and imperial terminology

Many PCB assemblers had difficulty in picking and placing small resistors such as 0603 (1608), without major investment in specialist equipment. Component manufacturers addressed this problem by introducing a resistor network or array concept, where several resistors are contained in one SMT package. This could be easily handled, and also achieved down sizing via function integration.

3. CAPACITORS

A capacitor is a component designed to store electrical energy. In its most basic structure, it consists of two metal plates separated by air, or an insulating layer referred to as the dielectric. The capacitance value C, is given by

$$C = KA/T \qquad (7.1)$$

C is in coulombs/volt (Farads), K is the relative dielectric constant 'Permittivity' (vacuum = 1), A is the area of overlay of the metal plates and T is the dielectric thickness. Class I dielectrics exhibit low K values but have excellent temperature stability. Class II and Class III dielectrics generally have lower K values but lower temperature stability. Capacitors can be classified as either fixed or variable, and electrostatic or electrolytic. The capacitance of the fixed capacitor remains essentially unchanged except for variations caused by dielectric material property changes brought about by variations in temperature. This characteristic is referred to as 'Temperature Coefficient of Capacitance' (TCC). Variable capacitors work on the principal of changing the overlap area of the plates, or the distance between the plates.

Electrostatic capacitors have solid material or air dielectrics and are electrically non-polarized i.e. the direction across the supply is not important. Electrolytic capacitors have either metal foil or solid slug anodes and are electrically polarized. The fixed value electrostatic 'Multilayer' ceramic capacitor (MLCC) is the most common component by volume in the electronics industry, and, therefore, potentially the source of most reliability issues during board assembly and use. MLCC failure analysis is discussed in section 4.1, Chapter 10.

3.1 Multilayer ceramic capacitors

The Multilayer ceramic capacitor is manufactured as stacks of dielectric layers separated by interleaved metallised electrodes, with the complete assembly encased in ceramic (Figure 7-4). In a typical manufacturing process, thick film silver-palladium capacitor electrodes are screen-printed on to wet (green, unfired) ceramic sheets (approximately 20 microns thick) using an interleaving pattern. These are stacked under pressure, dried, cut to size and sintered at a temperature around 1300°C. The electrodes must have a higher melting point, and platinum (1774°C) or palladium (1552°C) is normally used. For leaded MLCCs, the process is almost identical, except that connection leads are brought out from each termination. MLCCs are very rugged, with the electrode system totally enclosed and protected from the influence of the ambient atmosphere. Standard versions are able to withstand immersion in 250°C molten solder, as well as high humidity, without the need for further encapsulation. Future lead-free variants will be qualified to accommodate 260°C-soldering processes. MLCCs offer low inductance and resistance and a wide range of capacitance values from very low Pico Farads (pf), to 1 Micro Farad (μF).

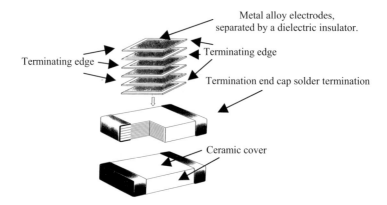

Figure 7-4. Construction of a multiplayer ceramic (chip) capacitor (MLCC)

Capacitors are classified by their capacitance stability across the operating temperature range i.e. temperature coefficient (TCC), which in turn is determined by the type and quality of the dielectric used to separate the metal plates. Figure 7-5 shows the variation between dielectric types with respect to capacitance change with temperature. The terms COG, NOP, X7R and Z5U are used to differentiate dielectric materials; the letters do not have any specific meaning. The temperature coefficient of COG (sometimes referred to as NOP) dielectric is close to zero, i.e. its capacitance value is almost constant over the full temperature range of the device. For other dielectrics, the TCC can be either positive or negative, or both, i.e. capacitance decreases or increases with rising operating temperature. The COG and X7R are more expensive and are used in many applications where Z5U versions would be inadequate.

Ceramic chip capacitors experience shelf life reduction of capacitor value over time. This ageing is primarily associated with Class II and Class III dielectrics and results in a reduction of capacitance value in time from the last heating (TOLH) of the device. During the early life of the capacitor, the random molecular structure of the dielectric slowly changes as individual dipoles align themselves in the same polarity, reducing the capability of the dielectric to hold a charge. With an X7R dielectric, it is not unusual to experience a 5% reduction of capacitance value during the 10,000 hours from TOLH and 15% for an Y5V dielectric material. Capacitors should remain in tolerance for 1000 hours after manufacturing date codes. When analysing Class II and Class III capacitors, the capacitance value may be greater than the upper tolerance when measured at less than 1000 hours from TOLH, and less than tolerance if measured in excess of 1000 hours from TOLH. Resetting a capacitor can be achieved by heating the device to 150°C for one hour to disrupt the dipole alignment and restore the material to its original state.

Surface mount MLCC case sizes are specified in the same way as resistors, i.e. 1206, 0402 etc. Although the actual height will vary as additional layers are introduced to increase the capacitance value.

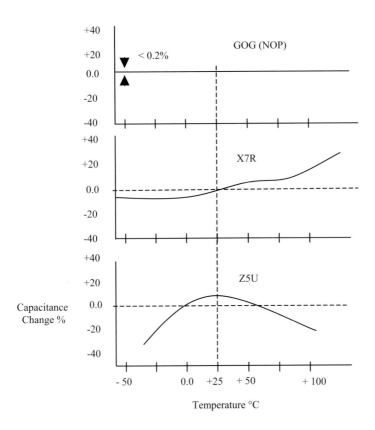

Figure 7-5. TCC graphs for three most popular dielectric materials, COG (NOP), X7R and Z5U.

3.2 Electrolytic Capacitors

For capacitance values greater than 1μF, a polarised electrolytic capacitor is required. These have a dielectric oxide film, which is grown electrolytically on a metal surface of either tantalum or aluminium. Such films have relatively low dielectric constants (k = 10 for aluminium, k = 25 for tantalum), but can be very thin. Wound electrolytic capacitors again depend on plate surface area, A, and separation, T, for the required capacitance value. They are constructed by interleaving two strips of etched aluminium foil between insulated separators soaked in an electrolyte. The strips and separators are stacked, wound and encased in an aluminium tube. External connections are taken from the electrodes to the outside terminals (Figure 7-6). A DC electric current is passed through the system to form a layer of oxide on the aluminium anode plate, and the oxide becomes the dielectric medium. The DC current must flow in one direction during the deposition of the oxide layer, and the positive terminal of current flow becomes the anode of the capacitor. Electrolytic capacitors are therefore intrinsically polarized, in that they need to be operated in a circuit where the applied voltage is of the correct polarity. Unlike non-polarized components i.e. resistors and MLCCs, the capacitor should be correctly orientated on the PCB with the most positive supply on the anode termination. Incorrect polarization failures are usually catastrophic and dangerous because the capacitor will explode under stress and become a short circuit. Due to this, electrolytic capacitors have been developed with an integral fuse to protect the assembly from reverse polarity capacitor damage.

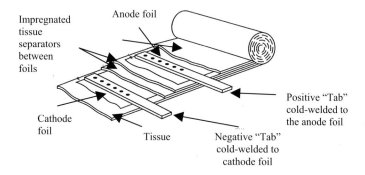

Figure 7-6. Basic construction of a polarised electrolytic capacitor

Increasingly, electrolytic capacitors are being manufactured with tabs underneath for surface mounting (Figure 7-7). These parts have the disadvantage of a relatively high profile and mass, and if not secured to the PCB, effectively, can cause unreliability from broken terminations (external or internal) due to vibration and lateral forces.

3.3. Tantalum Capacitors

Three types of tantalum electrolytic capacitor are in general use; wet foil, wet slug and dry slug (solid). Tantalum capacitors exhibit higher capacitance and voltage ratings than comparable sized aluminium electrolytic versions. They are also voltage polarised. Solid tantalum dielectrics are the most commonly used, and have been developed to meet surface mount assembly requirements. Solid tantalum capacitors, in miniature surface mount chip form cover values from 0.1μF to 100 μF and working voltages up to 50V. Capacitance value tolerances down to 5% can now be achieved. Each capacitor anode consists of high purity tantalum powder pressed and sintered around a tantalum wire. A tantalum pentoxide dielectric layer is deposited on the whole of the sponge surface, and contact to the top of the layer is made by impregnating the sponge with a solid electrolyte, such as manganese dioxide. The finished part may be dipped, or molded in plastic.

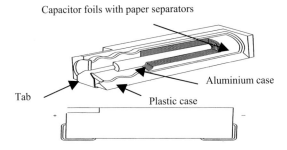

Capacitor foils with paper separators

Aluminium case

Tab

Plastic case

Figure 7-7. Construction of a surface mount, aluminium electrolytic capacitor.

Accidental reversal of electrical connection can cause damage to the equipment and operators. Several versions containing protection fuses are now available (Figure 7-8). Under reverse polarity, the dielectric suffers a total or partial breakdown depending on the magnitude of the reverse voltage. Similar consequences result where the voltage supply across the capacitor is of the correct polarity but in excess of the maximum forward voltage rating of the device.

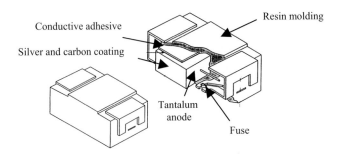

Resin molding

Conductive adhesive

Silver and carbon coating

Tantalum
anode

Fuse

Figure 7-8. Typical tantalum capacitor illustrating protection fuse.

The disadvantage of electrolytic capacitors is the higher leakage current, as compared to electrostatic capacitors, with the leakage current increasing with increasing temperature. This is caused by impurities in the foil and electrolyte, which also reduces the breakdown voltage. This phenomenon should be investigated during the failure analysis where continual breakdown of the dielectric is being experienced, when all other corrective measures have been unsuccessful.

4. TRANSISTORS AND DIODES

In transistor and diode package technology, epoxy resin has almost totally superseded metal and glass sealed devices. The topside of all silicon die (chips) are sealed and protected by a layer of glass, (passivation layer). Developments in silicon planar technology, improvements in passivation techniques (silicon nitride and derivatives) to protect the die surface, and advances in polymer formulations make it possible to mount silicon die on a free standing lead-frame, and encapsulate the complete assembly in epoxy resin by transfer mold techniques.

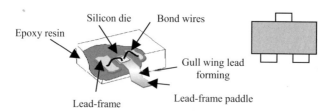

Figure 7-9. Philips style surface mount transistor (SMT). Code name SOT23.Exploded view and pictorial top view illustration

Packaging of surface mount and leaded devices is now automated, with die, lead-frame, molding and test, forming one production line. The Philips small outline transistor (later designated SOT23) with the Gull wing lead-forming format was designed with the lead frame exiting the device below the centreline of the package, with the die mounted on top of the lead-frame paddle. This is significant in the following sections, which discuss improved packages and miniaturisation.

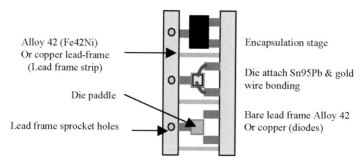

Figure 7-10. Basic sequence of SOT23 package construction. Bare lead-frame, die attached (die bonding) and wire bonding. Followed by encapsulation, solder plating and test.

The automated high volume production cycle of the SOT23 utilises a continuous assembly process, sometimes referred to as a Hoop Line. The basic manufacturing cycle is shown in Figure 7-10. The process takes the pre- electroplated alloy 42 (Fe-42Ni), or copper lead frame, and mounts the silicon die on to the lead-frame paddle using high melting point Sn-95Pb solder to prevent die/lead frame de-soldering during the PCB soldering process. Where a device has a complex lead-forming specification, the lead-frame material adopted would probably be a softer material such as tinned copper.

Some small low power devices now utilise silver-loaded epoxy instead of Sn-Pb solder to bond the die to the lead frame. After die bonding, gold bond wires (approx. 25μm diameter) are then ultrasonically bonded between the die base and emitter connections to the appropriate lead-frame tags. For a transistor, the collector connection is also made through the die base and the lead frame paddle. This is a major source of heat transfer from the device to the PCB tracking and heat-sinking. Encapsulation is achieved using epoxy resin injection molding.

After encapsulation but before separation from the lead frame, the component terminations are formed and solder plated using a dipping or electroplating technique. Figure 7-11 shows a continuous method of solder plating (dipping).

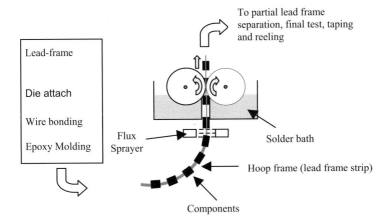

Figure 7-11. High volume, continuous solder plating process before component separation from the hoop frame.

Solder plating is a critical operation. Solderability is primarily a measure of the solder plating quality, or a measure of the oxide build up on the component termination through aging or poor storage. Oxide growth prevents solder flow and diffusion onto the termination during the PCB soldering process. Poor solderability is often referred to as a 'dry' joint because the lack of solder diffusion into the termination plating gives the appearance of a dry, dull joint. Termination solderability quality can have an serious impact on long term reliability, and many companies perform solderability audits on PCB and component terminations to track termination cleanliness. A standard method to assess solderability is the wetting balance test, which makes use of the meniscus layer that covers the surface all liquids (Figure 7-12). This is now a primary method to evaluate the solderability of plated surfaces. The IPC (Institute for the Packaging and

Interconnection of Electronic Circuits) has created an industry test standard IPC/EIA J-STD-002A Solderability Tests for Component Leads.

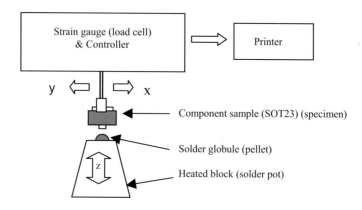

Figure 7-12. Basic blocks of the wetting balance solderability test system.

   The component or PCB section (specimen) to be analysed is suspended from a precision strain gauge (load cell), above a molten solder globule. The specimen is immersed into the solder to a specific depth and removed after a time. A force time graph is recorded which indicates the solderability quality (Figures 7-13 and 7-14).

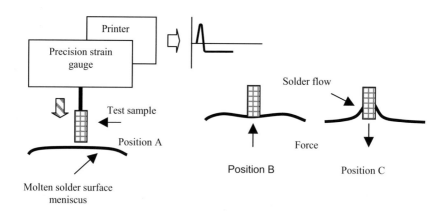

Figure 7-13.Qualitative measurement of termination solderability using molten solder surface meniscus.

When the specimen touches the solder surface (A), the solder surface meniscus exerts an upward force on it. The upward force, measured in micro Newtons (μN), increases to point B. As the termination heats and the solder wets the specimen, a downward force acts on the specimen pulling it into the solder globule, and the force falls from position B to C. The time to zero force is referred to as the wetting time. A wetting time of less than one second is typical of a clean oxide-free termination. Two seconds is considered as the industry standard maximum wetting time. Where a termination has a build up of oxidation, the trace will follow the line B to F where almost no wetting is experienced, or B to E where wetting occurs very slowly but outside the two second standard.

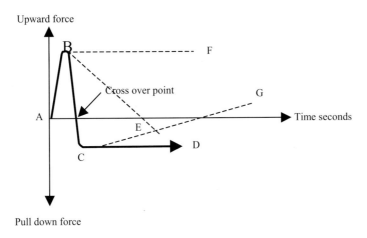

Figure 7-14. Component termination solderability force - time curve showing good and poor solderability examples.

Where plating is too thin or of poor quality, the initial wetting of the termination may look good, but as the plating dissolves into the solder or separates from the base metal, the termination will de-wet and follow the trace line G.

Traditionally, electronic component terminations have been plated with a tin-lead alloy of varying percentages from 40% to 5% lead. Recent (industry standard) lead-free termination plating solutions are, pure tin, tin/copper or tin/silver/copper alloys.

Immediately after the solder plating process, partial lead cropping, lead forming and final test operations follow to ensure that the component complies with the manufacturers functional specification. Partial lead forming separates the device terminations for test purposes, but one lead will remain connected to the lead frame strip to assist in automated testing and part number marking.

Due to their size, it is now impossible to mark most devices with their full part number. Each manufacturer uses a unique coding system to identify the part. For example, a transistor with a part number 2N1234 may be marked with the code B2, or a diode 1N123 with the code D3A. Generally, the code allocated to a specific component has no link to the alphanumeric format of the component part number. Manufacturers data books will list all part marking codes against the respective part numbers. Figure 7-15 shows examples of component marking and orientation bar where pin 1, (a) would be bottom left, counting anticlockwise to pin 6 in the case of (b) and (c) shows the manufacturing week code (week 35).

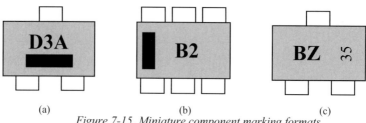

(a)   (b)   (c)

Figure 7-15. Miniature component marking formats

Finally, after part marking, the device is cropped from the lead-frame strip and passes through two further stages of functional test before being packaged in the transit packaging i.e. taped and reeled to an industry standard format i.e. IEC-60286-3 1998 (International Electronic Committee) or the component manufacturers bulk shipping packaging format.

4.1. Discrete diode and transistor packaging concepts

The Philips Small Outline transistor, SOT23, package is the European industry standard package profile for surface mount transistors. The Japanese component industry decided that it did not lend its self to miniaturisation. To accommodate future miniaturisation and retain compatibility with the SOT23, The Japanese industry standards associations (JIS) developed a new package, SC59. This component package, with the die mounted higher and under the lead frame, is promoted as having increased reliability because the longer terminations will absorb movement caused by thermal mismatch between the PCB and component. The construction of a transistor or diode package is basically the method illustrated in Figure 7-16, and 7-17.

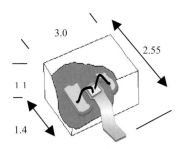

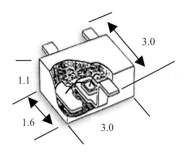

Figure 7-16. SOT23 package, with topside die bonding to termination paddle. Also shows gull wing lead forming. (Dimensions are in millimetres)

Figure 7-17. Inverted SC59 package illustrating longer lead lengths, die bonding to underside of lead frame and 90° lead forming (not gull wing)

Three terminal SOT23 and SC59 packages are used to package single diodes, which by their function only require two terminations i.e. Anode and Cathode. The first two-terminal surface mount diode package was basically a standard leaded packaged device, but with the leads replaced by end terminations. This metal electrode face bonding (MELF) package is used primarily in the automotive industry where the glass-sealed package gives excellent environmental protection from exhaust gases and humidity.

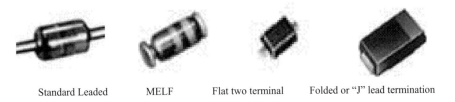

Standard Leaded   MELF   Flat two terminal   Folded or "J" lead termination

Figure 7-18. Two terminal diode packages developed from the leaded through hole technology to the modern flat package and fold-under "J" terminations.

Figure 7-18 shows the recent developments in diode packages. Package dimensions increase with component current and power rating but fundamentally they all utilise lead frame and bond wire technology. The key criteria are size reduction and number of components that can be placed per square area of PCB. The packages, shown in Figures 7.16 and 7.17, utilise bond wire connections, exhibit sensitivity to electrical overload and mechanical damage induced by thermal cycling. Some component bond wires are as thin as 25µm. The reliability of many package styles is determined by the mechanical and electrical durability of the internal bond wire connections. Rapid temperature ramps, mechanical vibration and component body flexing induced during PCB handling, often results in bond wire shearing. To address this, a patented sandwich style package has been developed Figure 7-19. The package has been designed for two-termination diodes devices, and comprises a silicon die sandwiched between the copper or alloy42 lead frame terminals. The improved durability to electrical overload and temperature cycling is due to the elimination of the bond wires, bond wire resistance and, therefore, reduced power loss and heat. The electrical and mechanical reliability of this package design has enabled the power rating capacity of devices to be up rated with no increase in package dimensions - a key feature for future miniaturisation.

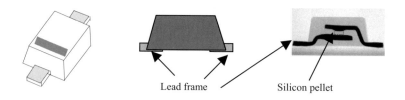

Lead frame   Silicon pellet

Figure 7-19. Sandwich construction diode showing basic package outline, profile and x-ray image of lead-frame and silicon chip (pellet)

4.2. Discrete Transistors and Diode packaging miniaturisation

Miniaturisation relates to the reduction in physical dimensions of electronic products. This goal must simultaneously be associated with the miniaturisation of electronic components. The advent of surface mount technology and the development of the SOT-23 and SC59 packages, described earlier, denoted the start of the process.

From the initial re-engineering of the three-pin surface mount transistor (SMT3) package, the internal construction concept did not change to achieve the ultra miniature (UMT) and extra miniature (EMT3) packages. Larger profile transistor and diode packages utilised Alloy42 (Fe-42Ni) lead frames. Copper began to dominate the lead frame material because it was easier to form and also had superior heat dissipation properties. Figure 7-20 shows an X-ray image of a typical lead frame format, which is used for SMT, UMT and EMT standard profile packages. EMT3, has pushed the boundaries of surface mount devices to a position where the circuit board area required for the component terminations is becoming significant with respect to the actual dimensions of the component epoxy body. To achieve and maintain zero defect targets during PCB assembly and to ensure electrical reliability in the field, sufficient space must be allowed between adjacent component connections to reduce the possibility of solder shorts and leakage induced by natural contamination.

The drive for continued component miniaturisation has led to the development of new packages in which the connections are contained within the periphery of the component body i.e. exiting from the underside of the component. The VMT3 (very miniature) package was the beginning of this trend. Figure 7-21 shows the dimensions of a VMT3 component and a photograph of the bottom side terminations. The terminations leave the component package flush with the base. The VMT3 package does utilise bond wire connection technology.

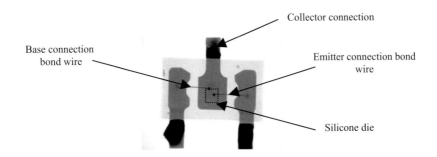

Figure 7-20. X-Ray image of a three terminal surface mount transistor. Lead-frame silicon die and bond wires are clearly identified.

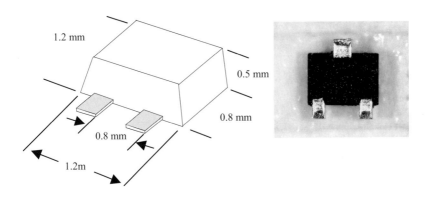

Figure 7-21. VMT3 package profile and bottom termination photograph.

This was the first package to require less than 2 square millimetres of PCB area. Considering that these devices are produced in very high volume (>10 million per month) and must achieve almost zero reliability failures in production to maintain profitability, this was an impressive achievement.

Down sizing trends of discrete component packaging are to provide an electrical function with the minimum requirement for PCB board space. The transition from SOT23 to VMT gives space reduction of 80%. To achieve further reduction, it became necessary to consider the area required for the terminations, as this became significant; with the EMT3 and VMT3 packages requiring 47% and 44% respectively of the total component footprint. The introduction of the chip scale package (CSP) required the development of new lead frames enabling external terminations to be contained within the maximum dimensions of the package plastic encapsulation.

Chip scale package technology for discrete transistors and diodes is a significant departure from the leaded package styles, discussed above. It builds on the established ball grid array (BGA) technology used for integrated circuits. The Chip Scale Package (Figure 7-22) is probably the smallest and lowest profile package that has been developed for discrete transistors, diodes and tantalum capacitors. Component manufacturers, introduced volume production of CSP devices during the first quarter of year 2002.

The trend in portable equipment towards miniaturised, low profile products will eventually push the acceptance of chip scale packaging at the expense of leaded devices. Table 7-2, illustrates the tremendous printed circuit board space saving that has been achieved from the surface mount SOT23 to the new CSP technologies.

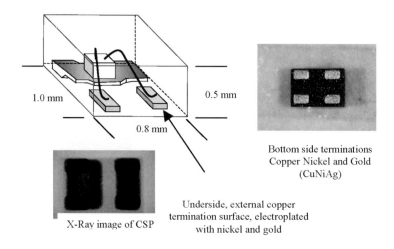

Figure 7-22. Chip Size package (CSP) construction. Basic exploded view, bottom side terminations and topside X-ray image.

The introduction of the CSP technology has achieved space saving of 89 percent over the SOT23 package format, compared to a power dissipation capability reduction of only 33 percent. This has been achieved via advances in packaging materials, improvements in silicon switching characteristics and the elimination of bond wire connections in diode packages

Table 7.2 Package dimension reduction against required PCB area.

Package	Dimensions (LxW mm)	Required PCB square mm	Space Saving percent of SOT23)
SOT23	2.5 x 3.0	7.5	Reference
UMT3	2.1 x 2.0	4.2	44
EMT3	1.6 x 1.6	2.56	66
VMT3	1.2 x 1.2	1.44	80
CSP	1.0 x 0.8	0.8	89

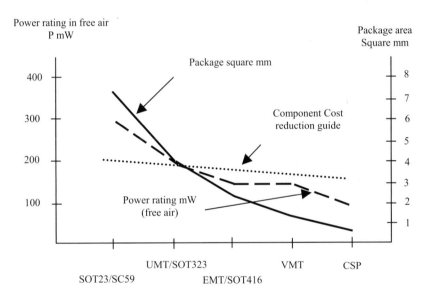

Figure 7-23. Package size reduction against package power rating milliwatts (mw)

Miniaturisation of components and silicon dimension reduces the power handling of discrete components. Fortunately, process developments in the switching characteristics of bipolar transistors and diodes have improved the overall efficiency of devices, enabling them to switch currents with reduced power wastage and hence lower heat dissipation. This is equally true for Metal Oxide Semiconductor (MOS) transistors, where the resistivity measured across its output when it is conducting, is a measure of the component electrical properties and its ability to conduct current with the minimum of heat dissipation. The power handling demands of most products are towards lower power consumption and longer battery life where the product is portable. Improvements in the silicon efficiency have enabled packages physically smaller than the SOT23, but with higher power ratings to be developed. Figure 7-23 illustrates the dramatic reduction in component dimensions against the more modest reduction in power rating of the respective package styles.

Minaturisation of discrete and passive components has the obvious advantages of space saving, product weight and reduced material costs, but also introduces the challenges of reduced tolerances throughout the assembly process to achieve build reliability. Miniaturisation also reduces the capability to pass circuit tracks under components. This a major disadvantage within the product miniaturization process. Figure 7-23 indicates component costs with respect to package size. The component costs have remained fairly constant with reduced material costs offset by increased manufacturing complexity and lower electrical and mechanical tolerance margins. Price fluctuations are ignored.

4.3. Component packaging and function integration

An alternative route to package miniaturisation is to increase the electrical functionality within a given package size. This strategy involves the integration of several components, for example, one or two transistors with integral passives components and diodes in one package. This reduces component placement count, and hence PCB assembly time and manufacturing cost. At the same time, the package is large enough to handle and place accurately on the PCB, with the minimum of expensive placement equipment. The down side is that the circuit designer has to work within the constraints of the functional configurations that are available to derive the required circuit function, and the PCB layout lacks the flexibility to place components where they are required. Using standard symbols for MOS, NPN, PNP transistors and diodes, Figure 7-24 illustrates examples of the many circuit configurations that are available.

With functional integration, or complex component packaging, the increased cost per component is offset by the reduced cost of placement. Assembly yield is more easily controlled with 0.65 pitch terminations as compared to 0.4mm, and the reduced number of external solder joints improves the assembly reliability. Also, PCB tracking can still be routed under the larger component package where, this is not possible with some miniature chip scale devices.

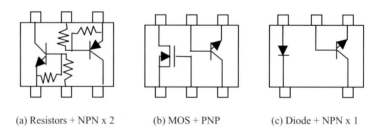

(a) Resistors + NPN x 2   (b) MOS + PNP   (c) Diode + NPN x 1

Figure 7-24. Illustration of typical functional integration within one package

Complex devices utilise two, or three, silicon die mounted on a single lead frame, with the die bond connections arranged according to the component function. Figure 7-25 illustrates a SMT5 (5 external connections) dual transistor. The image on the left shows a package acid etched to reveal the internal assembly. This technique is discussed in detail in Chapter 10, Section 3.

Functional integration has already begun to migrate into miniature packaging technology, i.e. dual transistors in EMT 5 and 6 pin packages, followed closely by VMT 5 and 6 pin styles. This gives the circuit and product designers the benefit of both functional electrical integration and mechanical miniaturisation. The complex device solution offers a serious option for product dimensional reduction, without the expenditure on specialist assembly equipment and process personnel training. Functional integration also provides the opportunity to increase product reliability, as integrated components exhibit a lower occurrence of failures, as compared with the overall reliability achieved by a collection of components mounted on a circuit board. The increased reliability is achieved by the total enclosure of the silicon die and associated parts within one hermetically sealed and thermally neutral environment, with respect to the thermal expansion coefficients of the materials chosen to construct the device.

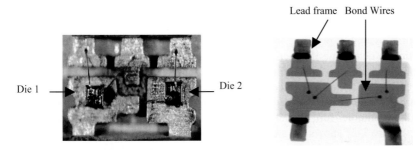

Figure 7-25. Open view of dual transistor package showing two transistor die and associated bond wires. Also shown is an X-Ray image showing lead frame and bond wires

Figure 7-26 demonstrates how the larger, more easily handled, complex package component can achieve an overall footprint reduction as compared to two miniature packages. The circuit board area and component spacing for the two EMT3 packages, is based on recommendations by the IPC Standards Organisation. Using a UMT6 package results in a 36% space saving. Where passives, as well as transistors, are incorporated in the device, typically 50 % of PCB area can be saved.

It is envisaged that the miniaturisation of discrete components will continue into 0201 body size and smaller, with the development of chip scale technology to reduce the termination space requirement. Experience is now being obtained in placing and soldering 0201 passives components.

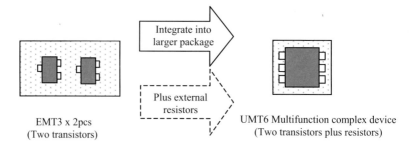

Figure 7-26. Comparison of required PCB area, with miniature components, against one larger complex multi function device.

5. LIGHT EMITTING DIODES

Although Light emitting devices (LED) are extremely reliable when operated within the manufacturers component specification, they are the cause of many reliability issues in the electronics assembly process. It is important to understand how to measure LED characteristics to determine if functional unreliability has been introduced during PCB assembly or during use. Both leaded and surface mount LEDs are used in high volumes, with the trend again towards surface mount devices as the light emitting intensity per chip unit area increases. Blue and white light emitting devices that pave the way for commercially available full colour moving images and displays have now been introduced. Figure 7-27 shows the structure of a typical LED lamp and versions of SMD (surface mount devices) in use today. Although the light emitting technology has changed, the mechanical construction of LEDs has remained predominately constant, other than the continual miniaturisation of surface mount device packaging. Leaded LEDs are often referred to as 'lamps' to differentiate them from surface mount devices.

The fundamental parameter of LED devices is their emitting colour, determined by the light-emitting wavelength of the silicon materials. Other parameters are, output viewing angle (directivity) and light intensity specified in millicandela (mcd). Colour is controllable within the LED manufacturing process; it is uncommon to change during PCB assembly. Viewing angle and light intensity are variable and linked to manufacturing process tolerances. For example, a red LED will be specified to have a wavelength of 650nm but its light output intensity (mcd) or brightness will vary between batches because of process tolerances. Because of this, production lots will be tested and ranked against the manufacturing company standard. Figure 7-28(b) shows typical histogram data where the target light intensity of the process is set between minimum and maximum levels. The process peak intensity of the part number, is designated the letter P millicandela by the manufacturing company. Devices falling into the intensity outputs of O, P or Q mcd will be accepted and ranked accordingly. LED intensity output is very important where many devices are used in close proximity to each other, for example in a running message board system.

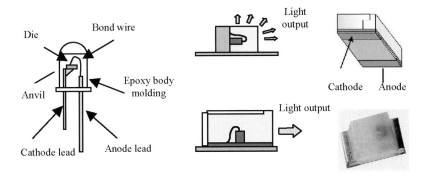

Figure 7-27. Standard LED lamp construction and single colour surface mount LED device.

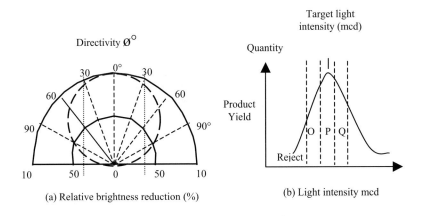

(a) Relative brightness reduction (%)

(b) Light intensity mcd

Figure 7-28. Light intensity viewing angle chart and light intensity ranking system.

Viewing angle or spatial radiation pattern is determined from the design of the LED lens Figure 7-28(a) shows a typical viewing angle chart indicating maximum brightness output, against viewing angle. Choice of lens is therefore important. From the viewing angle graph, maximum light output is achieved at angle zero (0°), viewing from 30°, would result in a reduction of perceived output by approximately 40%.

Light emitting devices have seen a major growth with the introduction of surface mount packages and, particularly, dual colour constructions. Figure 7-29 shows a dual colour device where two die are incorporated in one package - very similar to transistor or diode technology discussed earlier in this Chapter. A green die and a red die may be mounted in the same package, with cathode and diode for each die brought out separately to enable individual control of each LED. The device can, therefore, display red or green, or both colours giving an orange effect. Excellent effects can be produced where three or more LEDs are packaged together i.e. Red, Green and Blue.

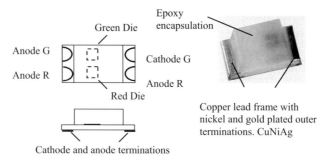

Figure 7-29. Dual surface mount LED incorporating two die

To ensure maximum light output clarity, most opto-electronic devices use an epoxy resin that does not provide a good barrier to humidity ingression. Where a light-emitting device has been stored in a high humidity environment (greater than 40% relative humidity), clouding may occur, as moisture that has been absorbed by the LED epoxy body turns to steam during soldering. Delamination of the die to epoxy interface, sheared bond wire, cracked die or LED body distortion can also occur. It is for this reason that surface mount LED manufacturers transport finished product (reels) enclosed in a sealed humidity proof envelope. After partial use of a reel, the remaining devices must be resealed. Any humidity ingression into the epoxy should be removed via an oven bake at 40°C for 48 hours.

6. PACKAGING OF INTEGRATED CIRCUITS

This section focuses on the construction, rather than the wide variety of specific device styles. Most devices use similar constructions and terminologies. As with the transistor families, users and applications have moved away from the ceramic glass sealed packages in favour of the high reliability plastic epoxies. Improved plastics and a silicon nitride passivation layer provide the silicon chip with adequate protection from encapsulation contaminates.

Surface mount dominates the integrated circuit packaging industry with Quad Flat Pack (QFP), Small Outline IC (SOIC), Ball Grid Array families and many more thin and shrunk variants of each. Figure 7-30 illustrates a typical dual-in-line (DIL) integrated circuit construction showing key parts, copper or Alloy 42 lead frame, gold bond wires, silicon chip and lead forming.

Surface mount, integrated circuit, packaging utilises the same construction philosophy as used for the leaded components, but with ever-decreasing mechanical dimensions and tolerances. Figure 7-31 (a) shows a typical 22-pin surface mount IC in SOIC format.

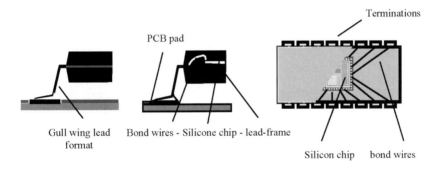

Figure 7-30.The construction and component parts of a leaded dual in line

Demand for increased circuit functionality on one silicon chip or die, and pressures to reduce the physical dimensions and weight of products, has rendered DIL style package, almost obsolete, in favour of the quad style package, with termination leads on all four sides. Increased functionality is normally associated with increased input /output (I/O) connections, making QFP and its variants very popular. The spacing or 'lead pitch' between terminations has decreased to 0.4 millimetres to offset the increased dimensions of the quad packages. This has affected PCB soldering process reliability, as the opportunity for solder shorts is increased.

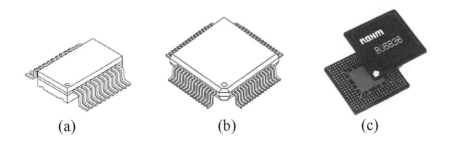

Figure 7-31. (a) 22 pin Small Out line Integrated Circuit. (SOIC), (d) 60 pin Quad Flat Package (QFP). (c) Ball Grid Array (BGA) package format.

Increases in package input-output connections have required development of new packaging concepts, which have utilised the area underneath the components as in chip scale technology (Figure 7-22). The lower surface of the integrated circuit incorporates a grid of solder, or gold balls, connected internally to the silicon die via bond wire technology. The concept is Ball Grid Array (BGA) packaging (Figure 7-32).

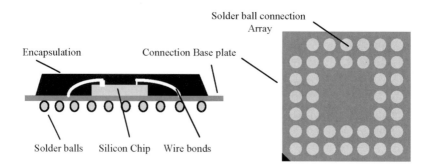

Figure 7-32. Typical ball grid array (BGA) package (only two bond wires are shown for clarity),
and base connection plate showing solder balls

QFP and BGA packages have enormous potential for high pin count but with much reduced mechanical tolerances with respect to the spacing between connections. Increased functionality, pin count, reduced lead pitch and, hence, mechanical exactness of component construction is resulting in greater collaboration between IC manufacturers and PCB assemblers, particularly with regard to high volume production and precision assembly. It is inevitable that failures will be more prevalent at this limit of current capability. Component failure analysis procedures must take into account the user limitations and materials employed, i.e. mismatch of thermal coefficient of expansion between fine pitch high density QFP and BGA packaged integrated circuits, and the printed circuit board material. The analysis must also ensure that the users soldering process temperature profile is in accordance with international standards recommendations. Chapter 10 reviews failure modes induced by the user and by the component manufacturer.

Section 9 discussed the integration of more than one transistor, diode or passive circuit component into one package and the benefits this conveyed. The integration concept has also been introduced to BGA technology, as BGA modules (Figure 7-33), and achieves a compromise between increased functionality per package and retaining component dimensions and tolerances within the capability of a wider population of users.

The internal construction of a BGA module is based on a high precision substrate system using alumina or high quality laminate printed circuit board. Components in die form, BGA format, chip scale components and possibly small BGA modules are mounted onto the substrate, wire bonded and encapsulated. The internal components of a BGA module are soldered to the substrate using a high temperature solder (300°C) to prevent them separating from the module during the user soldering process. The BGA module has the advantage of high-density functional capability within a very small area, but using a ball grid array pitch pattern, that is within the capability of most PCB assembly companies. The component manufacturers have contained the high precision assembly process within their control. Figure 7-33 also shows two types of solder ball connections. The standard eutectic solder ball collapses slightly during the pass through the soldering reflow process, effectively reducing the distance between the PCB and the BGA base. If this presents a problem, pillar type connections are available that guarantee a minimum underfill space. In underfill, an encapsulate material flows under the BGA, filling gaps between the ball connection to eliminate the ingression of contaminates and also increase the mechanical strength of the BGA attachment, This is particularly important in applications where the circuit board is subject to high acceleration, vibration or shear shock. Underfill is generally applied after X-ray or acoustic micro-imaging inspection (Chapter 9) of the quality of solder connections.

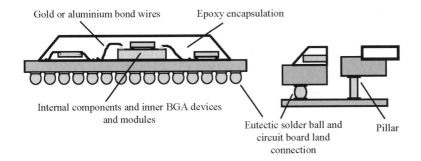

Figure 7-33. BGA module construction and two solder ball connection systems.

Electronic components are often marketed by function, such as digital signal processing (DSP), microprocessors, standard logic etc. Integrated circuits that perform memory functions, for example, dynamic random access memory (DRAM), static random access memory (SRAM), are unique in the sense that they contain larger than average silicon chips, but have relatively few input output connections. The memory I/O component will have Data in, Data out and a small amount of control functions. In these applications, the Thin Small Outline Package (TSOP) and its variants are used. The TSOP style package is a special variation of the small outline integrated circuit with connections on only two sides of the package. The TSOP has a very low profile, which is a key feature for mobile telephones. The package is also normally associated with fine pitch lead-out dimensions (0.4mm or less between each I/O pin). It requires knowledge and experience by the PCB assembler to handle, place and solder the IC accurately on the PCB to avoid mechanical and thermal stresses that may effect its long term reliability. Figure 7-34, illustrates a typical TSOP package and two variants that are available depending on the required PCB layout. TSOP I has the leads down the narrow edge and the TSOP II has the leads on the wide edge.

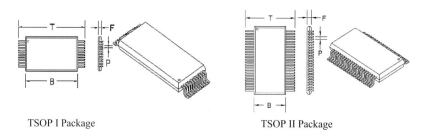

TSOP I Package   TSOP II Package

Figure 7-34. Integrated circuit packages showing two variants TSOP I & TSOP II

7. COMPONENT "BOND WIRE" CONNECTIONS.

Bond wire connections of any electronic device are a vulnerable point with respect to mechanical damage and component reliability. It is important to have a good understanding of the bond wire assembly process, as bond wire failures constitute a high proportion of IC failures - both user and component manufacturer induced.

Component lead frames are produced in strip form to facilitate automated production and mechanical accuracy, with the component cut and separated from the lead-frame at a later stage in the production process. The first stage in the manufacturing process is to attach the silicon die to the lead-frame paddle using high temperature solder (300°C), Sn-95Pb. This process is referred to as Die Bonding. Bond wires are then connected from each bond pad provided on the silicon die, and the appropriate lead-frame finger (Wire Bonding). Bond wires are predominately 25μm gold wires, although some manufactures use aluminium in less aggressive environments. Most modern equipment will produce between 20 to 25 bonds per second.

The ball-to-ball bonding system (Figure 7-35), feeds the Ag wire through the capillary feeder where the electric torch (spark) melts the wire end forming a ball. The ball is lowered on to the die bond pad, and a second spark welds the ball to the pad. The wire is cut and a second ball formed above the first. The bond wire is then pushed, and formed to position the second wire ball above the lead-frame finger, where it is attached using pressure and ultrasonic vibration. The formed bond wire profile is carefully controlled and designed to ensure no tension is introduced between both ends of the wire, and that the profile is maintained inside the package thickness tolerance.

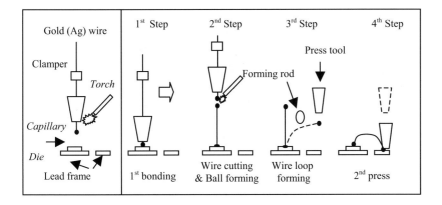

Figure 7-35. Ball to ball wire-bonding process

Figure 7-36 illustrates an alternative wire bonding process (stitch bonding), which is often utilised where a flame or spark is used to form both the die and lead frame bond attachments.

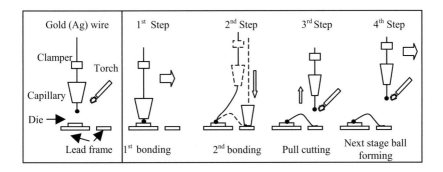

Figure 7-36. Ball to stitch wire-bonding process.

The choice of wire bonding process is generally determined by the component manufacturer's preference and package format, and to achieve maximum reliability and production throughput. Ball-to-stitch is regarded as current technology and more suitable for high volume component manufacturing. Scanning electron microscope images of ball and stitch bonding are shown in Figure 7-37.

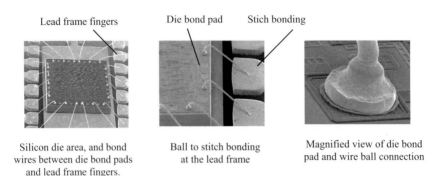

Figure 7-37 Scanning electron microscope images of typical IC pad bonding and close up views of ball to stitch wire bonding.

The completely bonded IC is then epoxy encapsulated via a transfer molding technique. The composition of the epoxy must comply with regulations concerning the restrictions of hazardous substances (RoHS) used in electronic components. Transfer epoxy molding utilises automatic inline processes where the component lead frame strip passes into the encapsulation process and the partially assembled components are clamped in a molding tool containing approximately 50 individual templates. Hot molding encapsulation material is forced into each template cavity. The process is repeated with the strip of components still attached to the lead frame. The next step is lead forming, followed by termination metal electroplating, functional test, separation from the lead-frame, final functional test, packing and shipping.

The storage and handling precautions of ICs in manufacturing, transport and user facilities are critical to minimize defects in the user manufacturing process. Where electrostatic discharge (ESD) damage is suspected (Chapter 10), a critical review of safe handling precautions should be conducted at all handling stations from the IC component manufacturer to the final circuit board assembler. With thin packages, humidity is catastrophic, as it turns to super-heated steam during the soldering process and causes package cracking, pop corning (craters on the device package surface) and delamination between the die and the epoxy encapsulation.

8. SUMMARY

There are numerous component packaging constructional configurations and electrical functions. This chapter has outlined the basics of component construction methods and identified issues that affect component and PCB assembly reliability. Chapters 8, 9 and 10 build on this knowledge and upon an appreciation of the reliability issues that arise during the component manufacturing process and in service.

CHAPTER 8

PRINTED CIRCUIT BOARD ASSEMBLY METHODS AND TECHNOLOGIES

1. INTRODUCTION

The modern printed circuit board (PCB) assembly process is extremely complex, highly automated, high speed and the source of many component failures. These failures can be introduced during the PCB assembly and the functional testing stage. This chapter aims to provide an understanding of the various process steps, and to create an awareness of the potential for failure or defect introduction during PCB assembly and test. Considering the high component volumes and the complexity of modern assembly lines, the occurrence of component failures originating during board assembly is extremely low. With the knowledge of component construction and packaging methods discussed in chapter 7, a good understanding of the circuit board assembly process is essential when analysing the cause of assembly-induced component failures.

2. PRINTED CIRCUIT BOARD DESIGNS

Fundamentally, there are two PCB design systems - through hole and surface mount. In through-hole process, components have connecting pins that are pushed through plated holes (vias) in the PCB and protrude through to the bottom side of the board. Generally, this system employs hand or wave soldering; where the assembled PCB bottom side is passed across and through a wave of solder which connects the protruding component pins to the copper circuit board tracking. Through-hole assembly has been replaced in many applications by surface mount circuit boards and components.

Surface mount PCBs do not have holes for component leads, but are designed with connecting pads on the surface of the board. The pads and components are interconnected to each other via copper or gold tracks. Surface mount components have connections that are almost parallel with the component body allowing the component to sit on the board surface, with the component connecting leads coinciding with the appropriate PCB connecting pads. Surface mount board assemblies are soldered using reflow soldering. Components may be mounted on the top and bottom sides of the PCB (double sided assembly) increasing the number of components to a given area of PCB. This in turn facilitates miniaturisation and reduced product dimensions.

Printed circuit board materials, for both types of board, utilise several materials depending on the application. The vast majority of applications use a laminate material referred to as FR4, which means "Flame Retardant" and the 4, is woven glass-reinforced epoxy resin. FR4 laminate is characterised by a reasonable thermal, electrical and dimensional stability during the PCB assembly and soldering process. The expansion coefficient of FR4 in the X-Y plane is approximately 16 ppm per °C.

3. COMPONENT STORAGE AND HANDLING

Component manufacturers issue documents, specifying recommended storage and handling precautions to reduce the probability of humidity and corrosive gas ingression. Recommendations are also given to avoid mechanical damage to component leads and to prevent electrostatic damage (ESD) to the device silicon chip. It is good practice to handle all components types within an ESD safe handling area, irrespective of technology.

Handling precautions differ from one type of component to another because of body size, lead pitch, component functional complexity and packaging material. The final stage of most assembly processes includes a PCB wash to remove solder residues. This wash may involve chemicals that may remove the component marking or even cause corrosion of the component encapsulation material and connecting lead plating. As the first step in any analysis, the failure analyst should ensure that the component user is complying with the manufacturers handling and storage recommendation.

4. COMPONENT PLACEMENT ONTO THE PRINTED CIRCUIT BOARD

The best-expected manual insertion rate of through-hole components is 200 to 400 components per hour. Automation has been a prime objective for printed circuit board assemblers since the electronics industry moved to manufacturing quantities that justified automated processes that ensure consistent quality and reduce manufacturing costs. There are areas where component reliability can be compromised during automatic handling and PCB insertion. Two formats of leaded-component taping are necessary for automated handling (Figure 8-1).

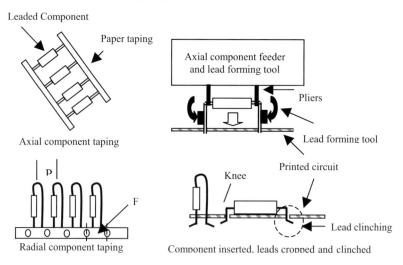

Figure 8-1. Illustrations of Radial, and axial leaded component taping, lead forming and PCB automated insertion.

The Axial format is more popular, and is supplied either in width, 'W', 52 mm or 26mm. The component leads are used as spokes and position locators for the automated insertion tool (component registration). When the component is in position, the pick-up pliers grip the component leads on either side and, simultaneously, the leads are cropped to separate the component from the paper tape. Arms are then lowered to form the leads into a 'goal-post' format to a pitch that meets the PCB plated through-hole pitch. The component is then lowered to the board with the leads passing through it. Some machines also include automatic lead cropping and clinching prior to the soldering process.

The radial component format is used where the board area is limited and the components placed upright on the PCB. Again, there are several radial taping formats, but the key parameters are component pitch, 'P', and lead pitch, 'F', typically 12.7 mm and 5mm respectively. Radial taping also utilises sprocket holes in the paper tape to ensure accurate component registration on the placement tool.

With both formats, it is important to ensure that the component body does not experience stress during lead forming and clinching operations. With the axial component, the lead forming or 'knee' must not be formed too close to the component body to ensure that mechanical stress is not transferred into the component, particularly in the case of glass body diodes. Clinching the leads too close to the PCB also introduces mechanical stress into the component.

Surface mount technology has accelerated the drive for high-speed automation to a position where it is now possible, on a medium volume placement machine, to place 20,000 to 30,000 components per hour, and on a high volume machine greater than 50,000 components per hour. The speed of placement is normally given as an average speed, as larger components take longer to pick and place compared with small discrete and passive components. To facilitate automation and high-speed component placement, the component must be delivered to the placement machine in a manner where it can be automatically offered to the equipment pick-up tool or nozzle. To achieve this, immediately after the final manufacturing and test stage of the component, it is incorporated into a tape and reel system. Figure 8-2, illustrates a taping system that is now industry standard and must comply with international standards with respect to dimensions, materials and mechanical durability. British Standard BS EN 60286-3 – 1998 (Packaging of components for automatic handling) is a taping standard, which is also matched with similar standards in the USA and Japan.

A surface mount tape and reel system comprises of a compressed polystyrene reel, a component carrier tape, either of laminated paper or polyester, depending on the component type and weight. Almost all resistors and capacitors will use paper tape carrier. Plastic carrier tape is generally referred to as 'embossed' tape. Finally, a polyester top cover tape is glued to the carrier tape to hold the component in the tape pocket. All parts of the taping system will be treated to be static dissipative with a resistivity of less than 10^{12} ohms. A static dissipative material will charge to some electrical voltage level if placed in an electrostatic field or rubbed against other materials (the tribo-electric effect). If the reel materials are inadvertently charged, they will discharge if given a discharge path. Being static dissipative does not mean that the materials will not charge. Manufacturers of PCB assemblies must take great care to minimise the possibilities for electrostatic charge build-up on the component packaging during handling and storage. Equipment should also be grounded in accordance with good ESD safe handling practices to eliminate any possibility of static voltage build up. Static can have a fatal effect on electronic components in the PCB assembly process. Static charge can cause miniature components to stick to the top tape as it is peeled back, presenting the component pickup nozzle with an empty carrier tape pocket and causing machine downtime.

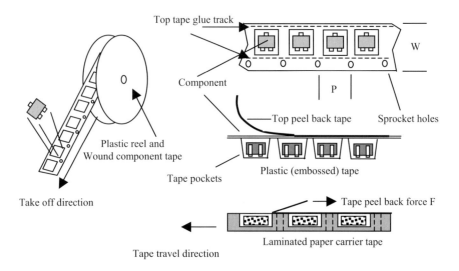

Figure 8-2. Typical surface mount tape and reel system holding several thousand components depending on reel size, component dimensions and weight.

Referring to Figure 8-2, the component reel is placed in a feeder mechanism, which has power driven sprocket pins that coincide with the tape sprocket holes (similar to a camera film drive system). As the tape is driven forward, the top tape is pulled back to reveal the component in the pocket ready for pick-up. British Standards specify dimensions for the tape width 'W', sprocket hole pitch 'P' and tape peel back force 'F'. Peel force may vary according to reel / tape age, storage conditions and feeder alignment. Peel force increases with storage at elevated temperatures. High peel force may cause tape breaking and tearing which again causes machine downtime. Peel force is specified at a peel back angle of 165° to 180° and pull back speed 300 mm per minute. It is measured for tape widths 0.8 and 56 mm, and is typically 1.0 and 1.3 Newtons, respectively.

5. SOLDER PASTING AND PRINTING (SURFACE MOUNT PROCESS)

To achieve accurate positioning of components, surface mount assembly PCBs must be designed to ensure that component pad (land pattern) dimensions and layout are in accordance with good practice and the component manufacturer's recommendation. For instance, where a component is using two pads, and one is larger than the other, the larger pad, will exert higher surface tension forces while in the liquid state and will pull the component off the smaller pad or raise it in the air on the larger pad. This failure mode is referred to as 'Tomb-stoning' because of the similar appearance to tombstoning. This problem also occurs where the heat sinking of one pad is much greater than the other, causing unequal solidification times for both ends of the component. Such failure modes are often regarded as a component solderability issue.

On a surface mount assembly, components are attached to the PCB using solder paste, which is a combination of solder balls (powder) and a compound containing binder flux and solvents, in roughly equal volumes. The consistency is similar to tooth paste. Solder balls are specified in many grades and sizes depending on the application and component lead pitch i.e. fine pitch 0.4 mm lead frame terminations would dictate a small dimension solder ball. Figure 8-3 shows an X-ray image of three solder paste samples, each consisting of different solder ball dimensions.

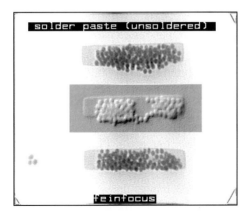

Figure 8-3. X-Ray image of solder paste showing metal solder balls suspended in flux, solvents and thixotropic agents. Image courtesy Fienfocus X-Ray equipment.

Solder paste may be applied to the PCB manually via a syringe at each individual pad or, more commonly automatically by stencil printing. Stencil paste application is significantly faster and is achieved using a metal stencil and squeegee blade. A metal stencil is produced with apertures that align exactly with the printed circuit board land pattern layout. The stencil is loaded with solder paste and positioned over the PCB and the solder paste forced through the apertures using a squeegee blade action. The angle of the blade is critical, and is adjusted depending on the printing density of the board. Figure 8-4 illustrates this process where the stencil, loaded with solder paste, is lowered and located accurately over the PCB. The squeegee is then pulled across the stencil forcing paste through the apertures and onto the component land pattern. The stencil and apparatus are then raised and ready for the next PCB to automatically line up below it. The stencil thickness and aperture dimensions, control the volume of paste deposited on the PCB

Stencil screen preparation and production is a high precision process, and is the only way to meet the printing tolerances required for fine lead pitch, chip scale and ball grid array (BGA) devices. High-density printing is normally associated with thinner paste deposits and finer solder paste metal (tin/lead) particles. Typical paste deposit thickness range from 150 microns to 250 microns.

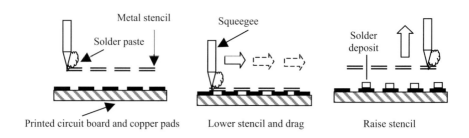

Figure 8-4. Basic printed circuit board solder pasting procedure using metal stencil and squeegee.

6. SINGLE SIDED SURFACE MOUNT PCB ASSEMBLY

Single sided PCB assembly requires components to be assembled on only one side of the circuit board and is the least complex of PCB surface mount assembly processes. Double sided and variants of surface mount and lead-through combinations are discussed in Sections 9 and 10.

After solder pasting, the PCB travels down the production line to the component pick and place station. Modern picking systems generally utilize air suction pick-up nozzles to collect the component from the reel or packaging format. Components are accurately placed on to the PCB using automatic vision viewing and alignment technology. Each component is placed on to the previously pasted PCB pads, where the tacky consistency of the solder paste secures the component in place during other processes of the board assembly.

Pick and place speeds can exceed 10 placements per second when the PCB contains many small components. Larger components require a slower pick and travel. Poor mechanical set up of the pick and place tool can cause cracked and damaged components if the component travels too far into the solder paste. Figure 8-6, illustrates a basic air pick up nozzle and single sided PCB with components placed ready for the solder heat stage.

Ideally, the PCB should pass through the solder heat process as soon as possible. Maximum acceptable dwell time is typically four hours. In free air, the solder paste deteriorates very quickly, causing unreliable PCB solderability. Also, solder paste remaining on the stencil for long periods, or solder paste containers left open, will cause solderability issues that are difficult to resolve, where operators are not conforming to company procedures. Again, this is a critical parameter when investigating component solderability failures.

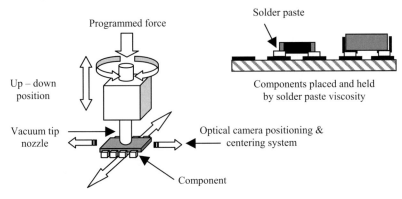

Figure 8-5 Program controlled component pick-up and placement head, showing pick-up nozzle movement positions and placed components held by solder paste viscosity

High-density pin-out, fine pitch components are generally placed after smaller components to avoid any movement of the fine pitch terminations due to vibration. Accurate alignment of all components is assured via automatic optical alignment (AOL) cameras, which also are programmed to reject components if they do not meet the pre-programmed outline or marking information.

7. RE-FLOW SOLDERING PROCESS

Reflow soldering is the making of solder joints by (re) melting previously applied solder, usually in the form of a paste. A re-flow machine is a conveyorised oven that transports the pasted and component-mounted PCB through a well-controlled heating system. This heats the solder paste, activates the flux suspended in the paste, and raises the temperature of the solder for a short duration (reflow zone) above the melting point of the metal powder. The liquid solder forms a set of soldered joints between the component connections and the PCB lands. After reflow, the PCB continues through a final cooling zone. The challenge in the reflow soldering process is ensuring that all components, terminations and PCB pads are brought to a re-flow temperature simultaneously, as each has a different heat absorbing capacity, or thermal mass. Smaller, darker coloured materials have a lower thermal mass than larger lighter ones and therefore, heat up more quickly. If the board is not brought up to the re-flow temperature in a gradual and controlled temperature gradient, components with a smaller thermal mass rapidly overheat and suffer permanent damage. The cool down process after the peak re-flow temperature is controlled, as too rapid cooling can also introduce component failures. A compromise exists here, as generally the faster the cooling; the stronger is the solder joint although the ductility may be impaired. A recommended cooling rate is 2 - 4°C per second.

Reflow ovens utilise conduction, radiation and convection to transfer heat onto the components, printed circuit board and solder paste. All have their advantages and disadvantages, but whatever the process used, the key parameters are the control of maximum component temperature and temperature gradients when heating and cooling.

All re-flow ovens take the boards through a series of heating compartments or zones; preheat, soak (activation), re-flow and cool down (Figure 8-6). The circulated heat is directed onto the topside of the PCB during its travel through the oven.

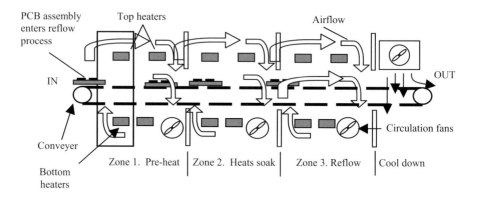

Figure 8-6. Forced air convection reflow oven heat tunnel with three heat zones and one cool down zone.

The PCB is passed though the tunnel at a constant speed, with the temperature of each zone precisely controlled. Temperature profiling is a key factor in the operation of a reflow machine to ensure consistent and high reliability solder joints. As part of component failure analysis, it is good practice to request a reflow profile from the machine operator to reveal the actual thermal history of the process. Temperature profiling is achieved by attaching thermocouples to a test PCB and connecting them to a data recording instrument. Temperature monitoring thermocouples must be fixed firmly to the component leads and PCB, because the circulating heated air will be approximately 30°C hotter than the monitored surfaces. A previously assembled PCB or test board is used to measure and store the time and temperatures of each thermocouple as the PCB passes through the oven. This process is repeated several times with adjustments to conveyor speed, heater energy and air flow until the desired profile is achieved. Profile settings are stored in the oven controller for specific PCBs and reactivated when required. Figure 8-7 shows a general target temperature profile shape for three heating zones, plus a cooling zone.

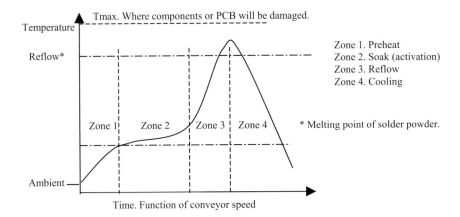

Figure 8-7. General target temperature profile shape for three heat zones and cool down zone.

Zone 1 preconditions the component and PCB, with a ramp rate of not greater than 4°C per second, to a maximum temperature of 150°C. Too rapid heating will thermally shock the components and PCB, causing delamination, bond wire stress/breaking and micro cracking of sensitive ceramic based components.

Zone 2 is the activation zone, since it activates the flux in the solder paste and allows unwanted binder and solvents to evaporate. It also allows time for temperature homogenisation.

The heating rate should be low if the size and colour of the components vary, but with uniformity, it may be 1 – 2 °C per second. If the soaking temperature is too high, premature evaporation and oxidation of the solder paste can impair solderability.

In Zone 3, the reflow zone, the paste is taken above the solder melting point allowing the component and PCB to form solder joints. The ramp rate from soak temperature to peak reflow again should not exceed 4°C per second. Maximum reflow temperature is set at approximately 25°C above the melting point of the solder ensuring that the component soldering specification, maximum temperature and time at maximum temperature are not exceeded, Most components are specified in a reflow process with an absolute maximum temperature of 260°C for 10 seconds, with two passes permitted. The reflow maximum temperature window is set by the component and PCB specifications, and the minimum is determined by the solder paste to produce a satisfactory joint.

Ideally in Zone 4, the cooling rate should mirror the heating rate. Forced cooling is preferred, rather than natural cooling to ambient. Again, there is a compromise between rapid cooling to optimise solder joint strength and reliability, and avoiding thermal shock.

8. WAVE (FLOW) SOLDERING PROCESS

Since the introduction of automated PCB assembly and component lead soldering, wave soldering has been the mainstream method of applying solder automatically to the bottom side of PCBs. Wave soldering utilises a stainless steel bath filled with molten solder, a solder heating element and an agitator to produce a stationary wave across the surface of the liquid solder. Considerable expertise is required to maintain the solder in prime condition, at the correct temperature and wave dynamics to achieve the best soldering results. Figure 8-8, shows the wave shape as the PCB passes over it

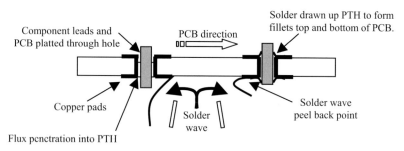

Figure 8-8. Typical leaded component and plated through hole (PTH) solder fillets achieved with the wave soldering process.

In the process, there are four stages; fluxing, preheat, solder wave and cool down. A fifth washing stage may follow if that is the quality assurance policy of the PCB assembler.

The main purpose of the flux is to clean and prepare the component leads, PCB plated through holes (PTH) and pads by removing oxide and compound films that inhibit the solder wetting process. Some fluxes are highly active and corrosive whereas others are very mild or mildly active. Grade "R" has very low activity, and grade "SRA" is very strongly active. Grade RMA is the most common and classed as mildly active. Strong fluxes tend to leave corrosive residues on the component and PCB, requiring costly washing and cleaning operations. Long term, these corrosive residues damage components and introduce unreliability of the product in later life. Flux can be applied in several ways, but spray fluxing is most common. Figure 8-9 illustrates a typical spray fluxing process.

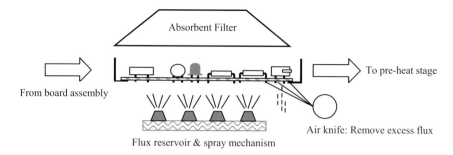

Figure 8-9. Example of PCB flux spray process prior to entering the preheat stage

Many systems incorporate an air knife after the flux application to remove excess flux and force flux inside the platted through holes. The preheat stage in wave soldering is very similar to that in surface mount production apart from the arrangement of the heaters, as shown in Figure 8-10. Preheating is achieved by blowing air through perforated heaters onto the underside of the board. The temperature of the PCB and components increases slowly until at the end of the preheat tunnel, the topside of the PCB and components are between 100 and 150°C. This reduces thermal shock when the PCB and components make contact with the molten solder wave. Thermal shock to the PCB in wave soldering is a serious concern. If the differential between the topside and bottom side is too great, the board will delaminate and warp, causing high mechanical stress on the components when the solder has solidified.   The preheat stage also activates the flux, so that all metal parts are clean prior to reaching the hot solder wave.

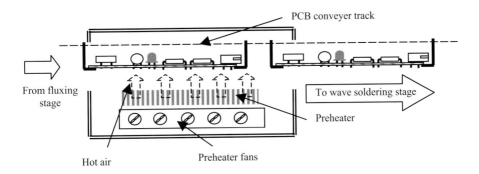

Figure 8-10. Preheater tunnel to raise the PCB temperature prior to wave soldering

The temperature profile through the complete process is shown Figure 8-11. The temperature drop as the board exits the preheat stage is caused, on some machines, by the difficulty in placing the wave immediately at the exit of the preheat tunnel. The temperature fall can be 20 - 40°C, which may be significant for large integrated circuits.

Attention to thermal profiling or any temperature excursion merits consideration for the component and its long term reliability. As described in Chapter 2, all physical dimensions of a ceramic component will expand or contract around 6 ppm per degree centigrade, whereas an FR4 board material will change by 14 ppm per degree centigrade. Due to these differences, the component terminations will be under tension when the solder solidifies and the completed assembly cools down. Components are usually designed for this but great care must be taken with the use of large components and the choice of PCB laminates, with respect to matching TCE. Such failure modes are associated with termination tear and solder joint fractures. The cool down stage in wave soldering systems maybe based on natural cooling to ambient or controlled.

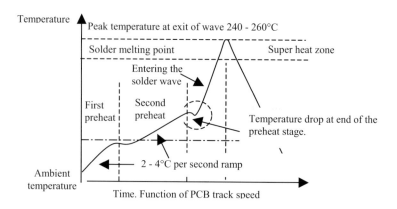

Figure 8-11. Wave soldering process, temperature profile trace

9. DOUBLE SIDED PCB ASSEMBLY AND SOLDERING

To achieve higher component density, the double-sided assembly technology has evolved. Double-sided circuit boards have copper tracking and component pads (land pattern) on the top and bottom side of the PCB, and often each side is interlinked by through-hole vias. Double-sided assembly requires the placement of components on both sides of the PCB (Figure 8-12). Where a double-sided assembly is totally surface mount, the solder pasting and component placement process is as described previously. To populate the underside with surface mount components, the PCB is inverted and the pasting and component placement process simply repeated. The second pass through the reflow oven does not melt the previously soldered side, as the joints are protected from the direct heat by being on the underside of the assembly.

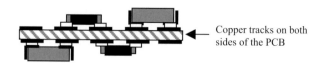

Figure 8-12. Doublesided surface mount printed board assembly

This method is very common, and almost all surface mount components are specified to withstand maximum reflow temperatures with two passes through the reflow oven. Further passes through any other soldering process may compromise the components reliability.

10. MIXED TECHNOLOGY PCB ASSEMBLY

There are still components that are not available in surface mount configuration (for example, some connectors and capacitors). Therefore, some assemblies have through hole and surface mount components mounted on the same PCB. Figure 8-13 shows a typical mixed technology board. The surface mount component is glued to the PCB with heat curable adhesive. Generally, the conventional components are inserted first and then the surface mount component adhesive applied before the board is passed through the pick and place line. The adhesive holds the component until cured.

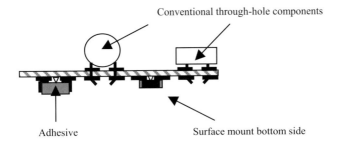

Figure 8-13. Mixed technology component PCB assembly.

No solder paste is used in this type of assembly. The surface mount components pass through the solder wave with the leaded-component leads, all surface mount components and protruding leaded component leads being soldered simultaneously (Figure 8-14).

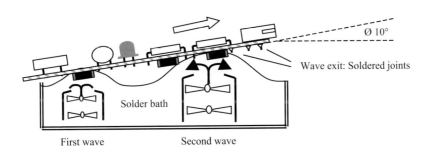

Figure 8-14. Double wave mixed technology wave soldering process.

After components have been inserted and glued to the PCB, the assembly passes through spray fluxing and preheat stages prior to entering the solder wave. Where components are mounted on the bottom side of the PCB, a dual-wave system is utilised to avoid unsoldered joints due to air pockets. The dual-wave system has a short wave, which circulates more violently than the main wave, dispelling air and trapped gases from the connections. The short wave is approximately 10 mm wide and the second main wave 30 to 40 mm wide. The component specification will identify maximum soldering temperature and duration for one pass only, as apposed to two passes in a reflow process. This requirement is often violated by PCB manufacturers, when a second pass is required, as in the case for a rework situation. Ceramic chip resistors are sensitive to additional solder passes, because the internal silver-palladium inner terminations will evaporate, causing partial or total open circuit. This is discussed in Chapter 9.

The dynamics of wave soldering are important. For example, to achieve the best solder joint and minimise solder shorts between component leads and PCB tracks, the relative speed between the wave and the component as it leaves the wave must be zero. Also, the board is inclined at about 10° to reduce the formation of solder icicles by using gravity to overcome the surface tension of the solder web or peel back point. (Figure 8-8)

Wave soldering of mixed technology boards is not ideal, particularly if there is a component mix on the board as shown in Figure 8-15. Surface mount and through-hole components are located topside and surface mount on the bottom side.

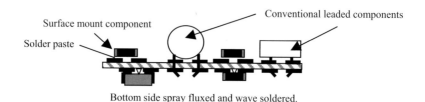

Figure 8-15 Double sided mixed technology PCB

With this arrangement, the topside surface mount components are solder pasted and the plated through holes for the leaded components are masked off to avoid solder blockage. Surface mount components are then placed on the solder paste and the PCB passed through the reflow process. The sequence to mount the other components varies depending on circumstances but generally the board is inverted, adhesive applied to the surface mount positions and components placed and adhesive cured. Through-hole components are then inserted in the topside through holes, and the bottom side of the board passed through the wave soldering process, as described previously. With very complex mixtures of components, or during rework, the PCB may be passed through the solder wave for a second time, which may compromise many components and long-term unreliability.

11. SINGLE SIDED MIXED TECHNOLOGY REFLOW PROCESS

A relatively new procedure has been developed to enable a single reflow operation to solder both surface mount and conventional through-hole components on the same board. This is particularly desirable since in wave soldering of very fine lead pitch components, it is becoming increasingly difficult to achieve zero defects with respect to solder shorts despite the use of solder masks. This soldering process is referred to as Pin-in-hole re-flow (PIHR) or intrusive reflow assembly (Figure 8-16).

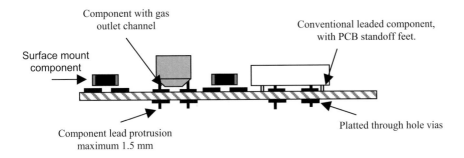

Figure 8-16. Pin–In–Hole–Reflow PCB with surface mount land pattern and conventional component through-hole vias.

The-pin-in-hole PCB design incorporates both surface mount component pads and plated through-holes. Despite the use of through-hole components, the process utilises only reflow soldering. The solder paste printing process is performed in the normal way (see section 8.5). Solder stencil thickness, squeegee speed and pressure are controlled to ensure that all surface mount component pads have the necessary volume of solder paste, and all plated through holes are at least 60 percent filled with solder paste.

After solder pasting, it is usual to place the surface mount components first and then insert the through-hole components. Through-hole devices are generally heavy and may be displaced during the automated surface mount component placement. Where through-hole components are hand soldered, care must be taken not to move the surface mount components or smudge the paste, causing solder balls to form on the PCB solder resist or mask (Figure 8-17). This may cause short-circuit connections between component terminations. Through-hole components selected for PIHR assembly should ideally have PCB standoff posts built into their body design, as shown in Figure 8-16. This feature enables the remaining solder paste, solvents and flux to gas out from under the component and permits the through-hole vias to fill with solder. The pasted and populated PCB is then passed through the reflow oven, with care again taken to ensure that the temperature profile is in accordance with the solder paste and component manufacturers recommendations.

With respect to component reliability in the PIHR process, the leaded-component body will be subjected to a reflow temperature of approximately 230°C, or greater, whereas in the wave soldering process, the top of the board does not experience temperatures above 180°C because it is shaded from the direct heat of the solder wave by the PCB material. Therefore, leaded through-hole used in a PIHR process must be selected to ensure that the body of the component can sustain the higher air temperature experienced in the reflow process. Connectors, capacitors and relays may fall into this category.

12. PRINTED CIRCUIT SOLDER MASK

A solder mask is a layer of protective material that completely covers both sides of circuit boards, except the copper areas to be soldered. The mask is applied by the manufacturer of the circuit board and will comply with the PCB assembler's specification. Solder masks are particularly important in wave soldering where the entire bottom side of the board is exposed to the molten solder wave. Without the mask, all bottom-side bare copper would be soldered, with a much greater potential for solder shorts between tracks. In reflow, the mask prevents solder paste migration between fine pitch tracks while the solder is in a liquid state.

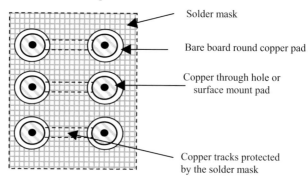

Figure 8-17. Printed circuit board, showing board and track areas protected by solder mask.

The solder mask is initially applied over the entire area of the board using dry powder vacuum laminators, which ensure that no air is trapped during the process. A previously prepared transparent artwork, with dark areas corresponding to the component pads and vias, is accurately placed on the PCB and the board exposed to ultraviolet light. The UV energy cures the film except the artwork protected pads and vias. Uncured, protected areas of the film are then removed by solvents, leaving a typical mask pattern as shown in Figure 8-17. In addition to preventing solder bridging between tracks and component fine pitch leads, the solder mask increases the PCB assembly reliability by improving its resistance to corrosive atmospheres and humidity.

Finally, for use in an exceptionally harsh environment, completed PCB and components are often covered with an additional transparent protective material called a 'Conformal Coating' applied by spray or dipping. Almost 100 percent of assemblies used in the automotive industry are covered with a conformal layer, with the thickness and quality selection dependant upon the protection level required by the application. A vehicle engine management PCB would have a high level of protection against humidity and corrosive sulphur dioxide exhaust gases. The aerospace and other high reliability industries are extensive users of conformal coatings.

GENERAL READING.

1 SMART Group web site *www.smartgroup.org*
2 Electronic Presentation Services (EPS) Bob Willis assembly process animation CD
3 Pin in hole reflow soldering. Electronic Production Journal, May 2001, T Weldon, Tecan Stencils.
4 British Standards BS EN60286-3. Packaging of Components for Automatic Handling

CHAPTER 9

FUNDAMENTALS OF FAILURE ANALYSIS

1. INTRODUCTION

1.1 Quality management systems

The industrial revolution signalled the advent of automated high volume manufacturing systems, with a series of formal controls to determine the outgoing quality level or product reliability. Quality control was seen as necessary to ensure that products met required minimum standards. A product that was inspected and judged not to be acceptable was scrapped, depending on the pressures at the time to meet customer orders. Quality was not directly linked to product design and production methods, nor was product lifetime reliability assessed. Quality was seen as the responsibility of the production line inspector.

Manufacturing and quality control methods introduced from the early 1960s completely reversed this philosophy. Whether to gain a competitive edge or just to survive, a company had to demonstrate that quality was their primary objective and the responsibility of the senior management and all employees - not simply that of the quality manager. The quality manager is now responsible for implementing a quality management system that encompasses company administration procedures, product design and manufacturing, with particular emphasis upon failures that occur during manufacturing or in service.

All failures, whether identified during quality inspection, functional testing or experienced by the customer, must be analysed, their root cause identified and effective measures taken to eliminate re-occurrence. This concept is referred to as Failure Mode Effective Analysis (FMEA) and is a key tool used within a design and manufacturing zero-defect program. Generally, two programs are in use; Design FMEA is used to correct failures associated with the functionality and construction of the component, and Process FMEA encompasses all aspects of product manufacturing, raw materials and assembly procedures.

To demonstrate that a company operates a formal quality system, it is necessary to develop all working practices according to a recognised quality standard, such as the International Standards Organisation (ISO) system that lays down the foundations for the company's administration, design and product manufacturing procedures.

1.2 Utilisation of failure data

Many manufacturing companies are implementing zero-defect programs, with a target quality level of zero manufacturing defects or functional failures at final test. In such a program, a manufacturing company assembling PCBs, for example, will collect and code defect data relating to assembly machine downtime, incorrect components placed on the

PCB, solder shorts or unsoldered joints, component functional failures identified during final test, and failures returned by the end user.

Using the zero-defect reporting programs, failures modes are entered into a database and plotted in failure code or categories. From this, the company will analyse the quantity of specific defect types and their individual impact, and work toward eliminating them. When component failures are identified during product functional test or field failure returns, they are analysed to determine the failure mode and the root cause of failure. The component manufacturer is usually requested to perform a comprehensive failure analysis of the component and issue an industry standard format failure report, which covers eight disciplines. The eight disciplines in this '8D' report are listed below.

Outline – Customer name, product part number, reference numbers, product quantity affected, product traceability data (if available) and complaint date.

Team members performing the analysis - Helpful for further communication and discussions.

Problem description - Specific product failure mode description. For example, no output, component open circuit, diode high reverse leakage current etc.

Containment Actions -These explain the analysis performed by the failure engineers. For example, optical or X-ray observations, functional-testing results, and visual images relating to the failure mode. Containment actions may be recommendations to quarantine similar components or finished products until a final analysis report is available.

Root Cause - Analyst observation and recommendations to the cause of the failure. Over-current or voltage, thermal cycling, PCB assembly damage etc.

Permanent corrective action - Where the cause of failure is the responsibility of the component manufacturer, the report will detail product recall and replacement actions, in parallel with production and process/procedure changes, and implementation dates to eliminate the failure mode. This would be implemented via the component manufacturer's zero-defect program. Where the failure mode is the responsibility of the component user, the analysis report gives as much information, via the root cause section, to enable elimination.

Verification of corrective action and effectiveness - Where the failure mode has a major impact on product quality, a component user or manufacturer will implement increased testing and auditing of the component, with particular focus on the failure mode and the effectiveness of the implemented corrective actions.

Actions taken to prevent re-occurrence - This section, if required, would be issued as an interim report discussing and committing to an agreed date to report on the long-term effectiveness of the implemented corrective actions. When the re-occurrence is deemed to be of an acceptable level, the failure report would be closed.

It is not uncommon for the report to confirm that no fault was found and that the component complied fully with the component manufacturers functional and mechanical specification. This will cause the manufacturing company, via its zero defect program, to examine and adjust product test procedures and parameters to eliminate this rogue failure mode and move closer towards zero-defects at final test.

The failure mode will be entered into the zero defect database and the results reviewed. For example, if the review shows an increase in component ESD damage from more than one supplier, then a full handling and ESD suppression program would be introduced to review all safe handling procedures in their ware-house, production line, and final test areas. If the ESD damage is being experienced with components only from one supplier, the company may request a copy of all handling procedure documentation from that component manufacturer.

Where a failure is the responsibility of the component manufacturer, that company will study the failure mode within their own quality and reliability zero-defect program to eliminate the defect, and report to the customer the cause of failure and the changes implemented to eliminate it.

Zero defects are recognised as being unrealistic in most manufacturing processes, but without a zero-defect program, supported by failure mode reporting systems, the level of outgoing quality is almost unknown. FMEA is a powerful tool, and with a commitment of adequate resources, a manufacturing company can restrict failures to a minimum, and in turn, improve product reliability. The most important aspect of quality failures is not so much that they occur, but knowing that they have occurred and taking corrective actions quickly.

2. DETERMINING THE CAUSE OF COMPONENT FAILURES.

2.1 Reliability of failure analysis procedure

Successful failure analysis of electronic components is very much based on knowledge of component construction and materials (Chapter 7), and of PCB assembly processes (Chapter 8). Also, reliable failure analysis can be achieved only with the correct equipment and tools, and the knowledge to interpret the results. The following sections describe equipment regularly used to analyse product defects, via high magnification or non-destructive internal examination.

The action taken to determine a failure mode, or its cause, depends on the item to be evaluated, e.g. the component for analysis may be delivered attached to a PCB, or it may have been removed from the PCB by the customer. It is good practice to encourage the customer to supply a PCB assembly containing the component for analysis. This allows the failure analyst to study the external and internal features of the component and solder joint area while the component is attached to the circuit board. Removing the component may introduce other unknowns or remove the original cause of failure e.g. a solder short, a component incorrect orientation or an open connection to the PCB. Prior to further examination, the component may be carefully removed from the PCB or kept attached but electrically isolated from all other components by cutting the appropriate circuit tracks.

Failure analysis should be conducted under very strict and well-constructed procedures. In many cases, only one component is available for analysis and measures must be taken to ensure that images and other observations are stored during each stage, to preserve the cause of failure during the initial stages.

2.2 Confirmation of component functional failure

The initial and most important step in the analysis program is to determine that the component has actually failed, or does not meet the manufacturer's functional specification. It is not unusual to receive components that comply fully with the manufacturer's mechanical and functional specifications. A report may have to be prepared that explains this condition, listing compliance with the component's major specified characteristics. Further discussion concerning the customer's testing procedures follows and modification made to the test equipment failure mode criteria.

Functional testing of passive components, such as resistors, capacitors and inductors, can be made using high-resolution multi-meters or Inductance, Capacitance and Resistance (LCR) meters. Most modern instruments are auto ranging and will quickly display, store and print out the required measurement. The analyst should study the operation manuals for each piece of equipment to ensure that no obscure measurements are made in the case of improper connection to the device under test (DUT). Some measured values will be very low, with the capacitance, inductance and resistance of the equipment leads and probes contributing significantly to the measured value. In these situations, the equipment manufacturer's measurement techniques must be adopted to compensate for such stray effects.

Functional testing of semiconductor components, or active components as opposed to passive components, requires additional laboratory equipment. Active components include transistors, diodes, optical light emitting devices (LEDs) and integrated circuits (ICs). A useful item of equipment is the Curve Tracer which electrically stimulates the device under test and displays the output characteristic curves as presented in the component manufacturers specifications against specific input voltage, current and signal frequency conditions. A curve tracer can generate many signal types of varying amplitude and energy, and is capable of generating very high voltages and currents. It is important that the analyst understands the operational procedure for this equipment since it is capable of destroying a good component, if not correctly set up. The stimulation energy of the curve tracer must initially be kept at a low level, and the resulting output trace compared with the manufacturer's data. Excessive stimulation energy on a device with an existing short circuit will result in high currents, which may melt and destroy the fault condition.

3. EXAMINATION OF ELECTRONIC COMPONENTS

3.1. Introduction

Instruments for the observation of small objects include the optical and the electron microscope, which may be employed in the scanning or transmission mode. The scanning electron microscope (SEM) is an essential tool in the component failure analysis laboratory, but the transmission instrument destroys the sample. For information about the internal condition of devices, X-ray and acoustic micro imaging are employed. This chapter outlines the principles of operation of instruments used in failure investigations.

3.2. Resolution and focus.

In any magnifying system, the image is accurately in focus only when the object is within the focal plane of the instrument. If part of the object lies above or below this, then that part of the image will appear out of focus. The range of positions on the object for which the eye can detect no change in sharpness of the image is known as the Depth of Field. In microscopes using light, this distance is small, and in order to produce sharp images, the sample must be very flat. This is a serious disadvantage if we wish to view three-dimensional objects, such as electronic components, at high magnification. The use of a beam of electrons brings an enormous improvement in achievable depth of field and the ability to resolve much smaller details of the sample (resolution). High-energy electrons have a much smaller wavelength, between (0.001 and 0.01 nm) than light (400 to 700 nm). Figure 9-1 compares the performance of an optical microscope at 700x magnification, and a scanning electron microscope image of the same object at 2000x magnification. Even with the increased magnification of the SEM, the difference in depth of field or focus range is striking. However, convenience of the optical microscope to obtain high quality images quickly, ensures that it remains an essential feature in any component failure analysis laboratory.

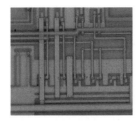   

Optical Microscope image, of the surface   SEM image, of the surface of a silicon chip.
of a silicon chip. Magnification 700X   Magnification 2000X

Figure 9-1. Magnified area of a silicon integrated circuit, showing aluminium metallisation connections. The optical microscope suffers from lack of depth of field as apposed to the SEM image with a large depth of field.

3.3 Electron beam interaction with specimen materials

In the SEM, a voltage, normally 30 kV or greater, accelerates electrons emitted from a hot tungsten source through a set of electromagnetic lenses that form the accelerating electrons into a very narrow beam less than 1μm in diameter. This beam is scanned repeatedly over the surface of the specimen in a similar manner to the raster pattern of a television tube. When an electron beam strikes the surface of a material, any of several events may occur (Figure 9-2). The electron beam penetrates the material to a depth depending on the beam energy and material structure. For most electronic component materials, the interaction depth is less than one micron. Within this interaction region, electrons in the material are scattered by the electron beam and some are scattered out of the material surface. Electrons, which are emitted from very close to the material surface are known as secondary electrons. Electrons emitted from sources a little deeper and more closely linked to the atomic structure of the specimen material are referred to as backscatter electrons. From deeper within the interaction envelope, X-rays are emitted with their energy and wavelength very much linked to the composition and atomic number of the material. The volume of backscatter and secondary electrons is associated with the topology (surface roughness) of the material. The SEM uses the secondary electrons to build an image of the sample surface. Backscatter electrons also provide information about the material composition, enabling identification of various phases or structures.

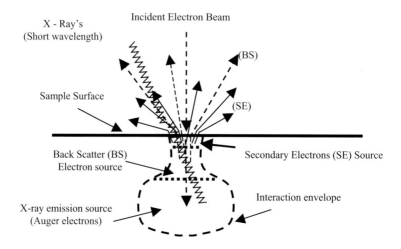

Figure 9-2. Interaction envelope, from which backscatter, secondary electrons and X-rays can be detected when a high-energy, electron beam strikes a solid object.

Figure 9-3 demonstrates how an image is constructed as the electron beam is scanned over the sample surface and electrons collected in discreet moments during the scan. Line-of-sight backscatter electron detection is very much influenced by the surface topology of the sample and provides the final black and white image with contrast and shade.

Figure 9-4 illustrates the complete system. Most modern SEMs utilise solid-state semiconductor detectors. The secondary electrons are drawn to and collected by the scintillator detector via the positive bias voltage of the detector grid. Once in the collector, they are accelerated to strike a phosphorus screen, and the energy is converted to light via the photo multiplier.

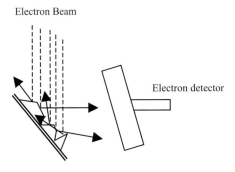

Figure 9-3. Schematic of backscatter electron line of site collection.

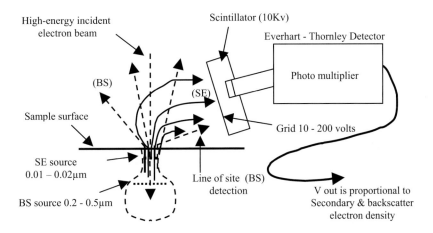

Figure 9-4. Representation, of an Everhart – Thornley secondary (SE) and backscatter electron (BS) detector. SEs are attracted into the detector by the v+ grid voltage.

Heating a tungsten- coated wire filament by an electric current causes electrons to be emitted from the tungsten surface (this process is referred to as secondary emission). The SEM uses a positive voltage to accelerate the emitted electrons through electromagnetic focusing and scanning coils and to strike the sample held at a high positive potential. The complete assembly (Figure 9-5) is evacuated to avoid scattering the electrons as they pass through the electromagnetic condenser and objective lenses forming beam less than 1μm in diameter. The electron beam repeatedly scans synchronously the sample surface. Emitted electrons are collected, converted to an electrical voltage and used to modulate the raster scan of the computer monitor display. The final resolution of the image is proportional to the electron beam diameter.

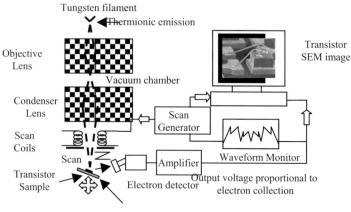

Figure 9-5. Scanning Electron Microscope, electron gun, signal amplification and scan/display block diagram.

Magnification is simply the monitor display area divided by the sample scan area. To enable higher magnification, the electron scanning area is reduced. Figure 9-5 shows the image generated when a transistor silicon chip is scanned. The chip die, bond wires and lead frame are clearly visible and demonstrate the excellent depth of field achieved with a SEM system.

3.4. SEM X-Ray spectral analysis

When the electron beam strikes the specimen (Figure 9-2), X-rays are emitted with a wavelength determined by the atomic number of the specimen material. Measurement of the wavelength enables the SEM to identify individual elements that are present. This process is known as spectral analysis, and gives information regarding the elements present and their quantity. During spectral analysis, the electron beam is stationary and not in the scan mode.

There are two techniques available to quantify X-ray emissions; wavelength dispersive spectrometry (WDS), and energy dispersive spectrometry (EDS), (Sometimes referred to as energy dispersive X-ray spectrometry EDX). In EDS, a silicon detector is positioned very close to the sample and inline of sight of any X-rays that are generated. Incoming X-rays excite a number of electrons into the conduction band leaving an identical number of positively charged holes in the outer electron shells. The number of electron-hole pairs is proportional to the X-ray photon being detected.

When a voltage is applied across the silicon chip, a current is induced as each X-ray is detected and absorbed. This current is amplified and stored in a channel of an X-ray analyser. EDS systems give reliable and rapid results for most failure analysis laboratory activities but lack in resolution and accuracy when detecting elements with atomic numbers below 14.

Wavelength dispersive spectrometry employs a crystal detector and diffraction grating to separate X-rays according to their wavelength. X-rays leaving the sample hit the grating separately and enter the detector in relation to their wavelength and grating position (tilt). Unlike the EDS system, the WDS detector does not have to discriminate between X-rays of differing energies, but simply counts X-ray photons arriving during a precise time window. Figure 9-6 shows a typical spectral trace of a component termination. From the trace, a high sulphur content is indicated which would cause termination failure by reacting with tin and silver, if it is present in the component termination.

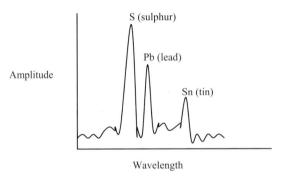

Figure 9-6. Typical trace obtained from an X-ray spectral analysis of a tin/lead solder joint.

In component failure analysis, when failure is suspected to be caused by a contaminant, spectral analysis of the failure area is essential to identify the contaminate, and determine the route of introduction. In the above example, the failed component was used in an automotive application. Sulphur ingress occurred via the vehicle exhaust system, despite the presence of a PCB conformal coating (see Chapter 8, Section 13), and entered the inner terminations of a surface mount chip resistor causing an open circuit of the component.

4. X-RAY IMAGING OF COMPONENTS AND JOINTS

4.1 Spectrum of electromagnetic energy and wavelength

Figure 9-7, illustrates the relative position of the X-ray energy spectrum with respect to long wave visible light and shortwave gamma rays. Gamma rays are generated from radioactive materials (isotopes) and have a shorter wavelength than X-rays; they are often used in portable X-ray inspection equipment where access to suitable power supplies is not available. The following section examines the generation of X-rays and their application to electronic component failure analysis. In the electronics industry, X-ray inspections can help to diagnose failures such as cracks or shorts in solder joints, and defects within electronic components caused by mechanical handling, excessive electrical currents or failure induced during the component manufacturing process. X-ray inspection is also used in production lines to develop high quality assembly processes that allow examination of hidden solder connections, such as ball-grid array joints on PCBs.

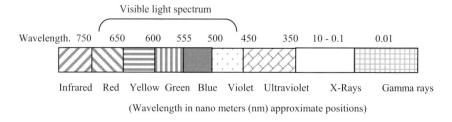

Figure 9-7. X-ray wavelength position related to visible light spectrum

4.2. Generation of X-rays

An evacuated electron gun is also used in the generation of X-rays, with the electron beam in stationary mode. A tungsten target material replaces the specimen. (Figure 9-8) and is positioned to maximise the quantity of X-ray emissions emitted from the electron gun window. The remainder are scattered and absorbed by the tube itself.

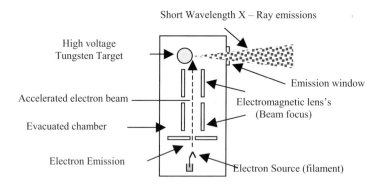

Figure 9-8. Basic construction and operation of an electron gun X-Ray generator.

As discussed in section 9.4, when a high-energy electron beam strikes a material several radiation energies, including X-rays, are emitted from the material surface. Accelerating voltages are generally 100Kvolts, or greater, and the high-energy electron generates light and heat, as well as emitted electron energy when it strikes the target material. Tungsten is ideal for this purpose, with a melting point of 2540 K and an atomic structure that emits a spectrum of short wavelength energies. The magnitude of the accelerating voltage controls the intensity of the emitted X-rays.

4.3. Detection and imaging of X-ray signals

Failure analysis laboratories use a real time X-ray imaging system, as opposed to X-ray systems that store the image on a photographic film. Real time image intensifier systems intercept the emitted X-rays, via a phosphorus photo cathode screen, and convert them into electrical signals that are amplified, digitised and stored. X-ray equipment designed for laboratory use must be completely sealed, lead lined and safe for use in an office environment.

Provision for remote position of the sample must be provided via the central control panel. The term 'real-time' imaging means that the X-ray image is continually changing as the sample is rotated and positioned within the X-ray emissions. This is a powerful feature when investigating electronic components for defects. Figure 9-9, shows a magnified image of bond wires of an integrated circuit.

When irradiated, the sample effectively casts a shadow onto the image intensifier screen. X-ray energy is absorbed by the component parts in differing degrees, depending on the individual material density or atomic number. For example, the epoxy encapsulation and silicon of an electronic component have relatively low atomic numbers and absorb a minimum of radiation, whereas the copper lead-frame and the lead in the tin-lead solder have atomic numbers of 29 and 82 respectively. Lead absorbs a high percentage of the X-ray energy and will cast a strong shadow on the screen, which will be imaged as black. The copper will absorb less and will be imaged as grey. To optimise image contrast, the operator controls the X-ray emission intensity by varying the accelerating voltage. To study a silicon chip, exhibiting low X-ray energy absorption, the X-ray emissions would be set to a very low level to ensure that the limited absorption would still produce a shadow.

A typical system will give magnifications up to 100x. This is more than adequate to study electronic component bond wires, silicon area, and internal construction for potential defects. Changes in magnification are simply achieved by altering the distance between the sample and the X-ray emission source.

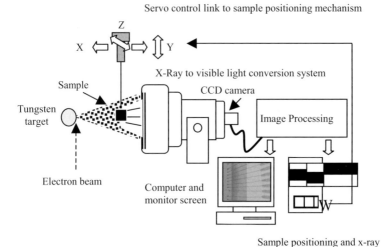

Figure 9-9. Real- time, radiography X-ray inspection system. Integrated circuit bond wire examination

5. ACOUSTIC MICRO IMAGING (AMI)

Ultrasound is generated via a transducer emitting vibrations into and through a liquid carrier. The echo or 'through transmission' of the ultrasound is detected and amplified by a receiver. The basic concept of this technology makes use of the fact that ultrasound travels with the minimum of attenuation through a liquid, but will be attenuated to varying degrees by solids and voids. The electronics industry has applied this technology in the internal examination of components for the detection of voids, cracks and delamination. More recently, it has been applied to volume screening of components in a production or goods inward environment.

Ultrasound frequencies in a range from 10 to 230 MHz. are used. High frequencies are required for increased resolution (short wavelength) and lower frequencies are needed to penetrate thick and energy absorbing materials. The use of AMI in the component failure analysis laboratory is now almost universal. AMI technology is a complementary technique to X-ray examination, and particularly useful for detection of micro voids, cracks and delamination. Like X-ray technology, AMI is non destructive, and can be applied as the first line of examination of components while they either remain mounted on a circuit board or loose, prior to performing destructive sectioning or package opening (see Chapter 10).

Since high frequency ultrasounds do not propagate well in air or vacuum, AMI is particularly suited to detection of unintentional air gaps, voids between and within materials, cracks and porosity. Maximum signal attenuation is obtained when such interfaces are encountered in a solid object. Figure 9-10 illustrates some of the defects in an integrated circuit.

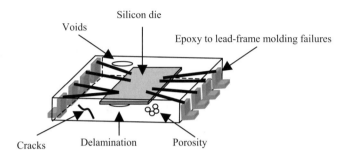

Figure 9-10. Integrated circuit package illustrating defect types effectively detected by acoustic micro imaging AMI.

There are several AMI configuration variants depending on the specific failure mode being analysed. Two systems that conveniently explain the bases of the technology are the scanning laser acoustic microscope (SLAM) and the C- scanning acoustic microscope (C-SAM). SLAM imaging is often referred to as through scanning as it detects defects by analysing the acoustic energy after it has passed through the sample (Figure 9-11). SLAM equipment utilises an ultrasound transducer that emits a continuous sound wave through a coupling liquid and the sample, and onto a laser scanning receiver assembly (a laser microphone). The microphone comprises a transparent plastic detector a 'coverslip' the underside of which has a very thin metallic mirror plated onto its entire surface. The need to immerse the sample, transducers and receiver in a liquid is seen as a disadvantage of AMI technology, particularly in a volume production-screening environment, but recent advances in inline high-speed drying techniques have lessened this concern.

The SLAM scans a narrow laser beam across the surface of the coverslip making approximately 30 scans per second. This laser beam is reflected back from the flat metallic surface at a known deflection angle. When the coverslip metallic surface absorbs the incident ultrasound energy, it ripples and the laser beam is reflected back at a different angle. Where the ultrasound is attenuated, or masked from the metal film, by voids and cracks in the sample, it will remain flat and the deflection angle of the laser beam is unaffected.

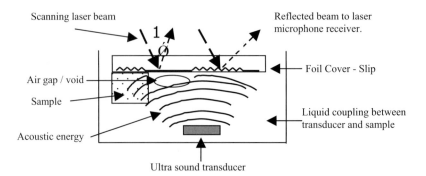

Figure 9-11. Scanning Laser Acoustic Microscope (SLAM). Acoustic wave is attenuated by the void in the solid material.

The SLAM detects discontinuities in a bulk sample by casting a defect sound shadow onto the coverslip laser microphone, modulating the reflected laser beam and displaying an image of the discontinuity. As sound travels through solids and voids depending on the material density and the signal frequency, the transmitted ultrasound frequency is selected to give the best image resolution. SLAM makes an ideal tool for high volume sample testing where only the existence of discontinuities is required but not their precise position with respect to depth. Figure 9-12 illustrates a black on white image generated with a SLAM instrument of a multilayer ceramic chip capacitor (MLCC). Where no discontinuity exists, the display will appear totally white, except for some grey imperfections that are a characteristic lack of definition and focus of SLAM imaging. The MLCC image (Figure 9-12) highlights where a void exists between the inner metallisation plates and the capacitor dielectric. This type of defect has now been almost eliminated due to improved dielectric materials and baking processes.

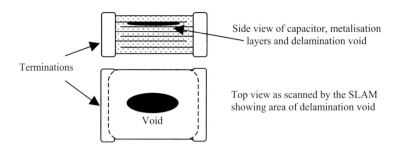

Figure 9-12. SLAM imaging of a multilayer chip capacitor exhibiting serious delamination and voiding between metallisation and the dielectric material.

A shortcoming of the method is its lack in image detail, with respect to definition and resolution. Greatly improved defect definition and location is possible with the C-mode scanning acoustic microscope (C-SAM). This also utilises liquid coupling to the sample and an ultra sound transducer, which is focused, pulsed and scanned across the surface of the bulk sample. The advantage of the C-SAM system is that an adjustable gating system is used that listens for echoes at specific time intervals from the precise time that the sound pulse was transmitted (Figure 9-13). This enables the system to develop an echo image within the depth of the sample as the gating time is increased until the maximum time is reached for the ultrasound to travel through the sample.

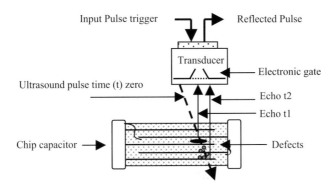

Figure 9-13. C-SAM pulse – echo ultrasound defect detection system, showing sound pulse, and echo time gating process.

At set time intervals, the detector gate is opened to listen for echoes at cross sections within the sample depth. Figure 9-14 illustrates this process as the gate signal sample time slot is effectively swept down through the internal cross sections of the bulk sample.

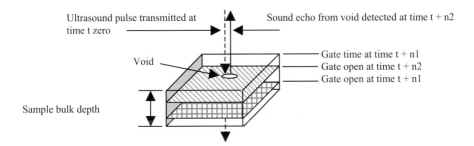

Figure 9-14. The echo detection process showing how infinite control of gate open time forms an image of the volume of the defect by cross sectional area and depth.

The C-SAM utilises the fact that sound waves propagate through materials at velocities determined by material density. The polarity of the echo signal changes when the sound wave passes across the interface between a solid and a vacuum, and when it passes from a vacuum back into a solid, i.e. when the sound passes through a void or crack interface. The echo signal is processed depending on its polarity to build a three dimensional image of the defect.

Where the C-SAM technique is used in a through scan-screening mode, the echo gate is continually open and the echo depth is not registered. The C-SAM defect image will be white on black i.e. where no defect exists the image will be black. C-SAM imaging resolution is greatly improved as compared to SLAM, but is considerably slower. When the echo-gating feature is utilised, and to differentiate between voids and defects that overlap each other, most C-SAM equipment will allow the imaging process to store and display echoes in different colours corresponding to varying depth or time zones.

Acoustic micro imaging (AMI) has developed into a very powerful tool. Many standards organisations are now specifying the need for AMI inspection in high density multilayer printed circuit board designs, incorporating encapsulated modules, high pin count ball grid array, flip chip and chip scale devices (see Chapter 7).

6. SUMMARY

Optical microscopes are fundamental to any failure analysis laboratory being easy to use and understand. However, most professional electronic component failure analysis laboratories employ more specialised and sophisticated equipment which utilise electrons, X-rays and sound to explore the surface and internal structure of electronic components. The level of complexity is determined by the scope of the laboratory and the budget. In many companies, it is often necessary to design and build customised variants or test platforms for effective product failure analysis and to ensure maximum product reliability monitoring.

RECOMMENDED READING:

1   P.J Goodhew and F.J Humphreys, Electron Microscopy and Analysis, Taylor & Francis Inc. 1988.
2   Perry L Martin, Electronic Failure Analysis Handbook, McGraw Hill, 1999
3   Mortinmer Abramowitz, Microscope Basics and Beyond, Olympus Microscopes, 1985.

CHAPTER 10

ANALYSIS OF PCB AND ELECTRONIC COMPONENT FAILURE

1. INTRODUCTION

The electronics industry encompasses numerous disciplines to produce a working assembly. These include;

Materials science of component raw materials,
Component design and construction technology,
Quality control testing for component reliability and durability,
Production batch and raw material traceability techniques,
Packaging and safe handling of components in transist,
Electrical functionality and specification understanding,
Bare PCB design and manufacturing process awareness,
PCB assembly processing,
Solder processes and science,
Component pick and placement systems,
Automated circuit board testing and test program philosophies,
Understanding product design and potential failure modes,
Failure analysis tools, and the science of failure analysis.

Component failure analysis is an engineering science that requires an understanding of all disciplines listed above, together with an open and critical approach, so that the final report is both accurate and constructive. Previous chapters have introduced these disciplines and their relevance. Using this knowledge, the present chapter describes failure analysis techniques with examples, and the necessary ingredients of a failure report.

2. COMPONENT PREPARATION

Successful component failure analysis involves diagnostic engineering skills that ensure that the component survives the analysis long enough for the failure mode to be identified, notes taken and electronic images stored. Some aspects of failure analysis involve procedures that can destroy the component and obliterate the failure mode before it has been investigated completely. In contrast, component functional testing, X-ray and acoustic micro imaging equipments (AMI), are non-destructive. i.e. they have no impact on the component construction or modify the failure mode. However, the component generally requires preparation for subsequent internal high magnification examination, to provide additional information and confirm the failure mode identified during non-destructive analysis. The chart below represents a typical analysis flow diagram.

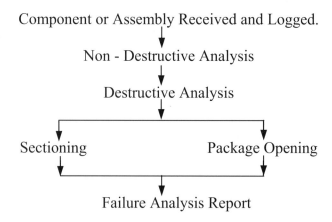

Component or Assembly Received and Logged.

Non - Destructive Analysis

Destructive Analysis

Sectioning   Package Opening

Failure Analysis Report

Sectioning and package opening (decapsulation) are destructive procedures that allow examination of the component in various cross-sections with the component epoxy encapsulation, partially or totally removed. The following sections examine these stages in more detail.

2.1 Component sectioning

Sectioning enables physical examination of component failure modes. The process is regarded as destructive analysis, because the electrical function of the component is usually destroyed. With great care, if required, it is possible to de-capsulate some components and maintain full functional performance for direct die probing, via the metallisation layers of the device.

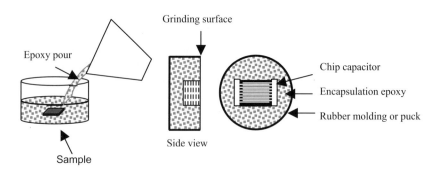

Figure 10-1. Component encapsulation and sectioning for internal examination.

Sectioning, involves encapsulation of the component in epoxy to facilitate handling and to support it during the sectioning process. Figure 10-1 illustrates the encapsulation of a multilayer chip capacitor. Using a rubber mold, the component is immersed in a fast-setting epoxy, and subsequently ground down to the desired depth. The location of the failure may have been determined from the initial non-destructive analysis. If not, more careful and gradual sectioning is required.

The epoxy should provide support to the delicate component edge, if the area of examination is close to the encapsulation interface with the component. The molding compound must exhibit high hardness, excellent edge retention and low shrinkage, and should cure at ambient temperature. Epoxies comprising of a resin and hardener are generally suitable for sectioning electronic components. They provide excellent edge retention and good support to the sample outer edge due to their high viscosity in liquid form. Epoxy cure times are generally 30 minutes and should exhibit minimum amounts of shrinkage during curing. Transparency of the epoxy is not often a criterion for electronic component sectioning.

Since only one sample is available and destructive analysis is irreversible, it is essential that the encapsulation process must be conducted exactly to the epoxy manufacture's instructions with respect to mixture composition and curing conditions. Defects, such as air bubbles, poor edge retention or soft mold may result. A failed mold can be recovered by acid etching of the encapsulated component and the process repeated, but the additional process step will increase the risk of destroying the failure defect.

Successful encapsulation prior to sectioning depends upon the cleanliness of the sample. Finger oil and dust can severely degrade the adhesion of the epoxy to the surface of the sample. Typical sample cleaning, prior to encapsulation, entails a mild detergent and water wash, followed by an isopropyl alcohol rinse and thorough drying using low-pressure warm airflow. More aggressive cleaning may be accomplished using an ultrasonic cleaner, although care is required, as vibrations can damage integrated circuit bond wires or attachment to the lead frame or silicon bond pad.

Where a defect is very small or close to the component surface, vacuum encapsulation of the sample maybe advisable. This extracts any microscopic air bubbles trapped in the contours of the component or in the epoxy that may prevent complete adhesion between the epoxy and the component surface. The level of vacuum is not critical and requires that the internal pressure be reduced by approximately 50 percent. Immediately after mounting, the sample is placed into a bell jar, which is evacuated slowly until the epoxy starts to froth. At this point, air is slowly returned into the chamber. It is good practice to repeat the process immediately to ensure efficient air removal. With large components, the mold can be partially filled and evacuated in stages.

Most encapsulation failures are caused by poor cleaning, use of overaged epoxy or incorrect mixing. Soft epoxy or epoxy pull-away occurs when the epoxy fails to cure completely and adhere to the sample surface. The problem may be remedied by oven heating at 100°C for approximately 30 minutes to accelerate curing. Re-encapsulation or back filling may rectify pull away. Often, the preferred solution is to re-capsulate, i.e. grind away the bulk epoxy and submerge the sample in dilute nitric acid until the epoxy is removed. The process is then repeated.

Before commencing the grinding procedure, the epoxy mounts are removed from their molds and identification codes applied accordingly to the local tracability procedures of the analysis laboratory. The sharp meniscus around the top surface of the mount must be bevelled off to prevent cutting and tearing the grinding paper used in the initial grinding steps.   The final, and most critical, step is to mark the plane of sectioning, and to approach the failure area in a direction determined from prior X-ray or acoustic micro imaging analysis. If not identified correctly, the structure of interest can easily be ground away and lost.

Sample grinding is performed by holding the sample against a rotating platen covered by a paper disc impregnated with silicon carbide particles (Figure 10-2). Throughout the process, a lubricant, usually water, is flowed onto the face of the abrasive paper to control heating between the abrasive and the sample, and to remove the grinding deposits. Despite modern technology and mechanical ingenuity, sectioning remains a manual craft, with personal preference and techniques based on experience.

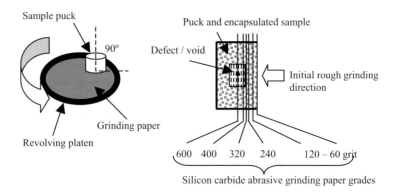

Figure 10-2. The grinding wheel and grinding abrasive sequence from rough grinding to sample surface polishing.

The sectioning operation involves grinding at diminishing levels of roughness, rotating the sample through 90^0 at each transition (as for metallurgical specimen preparation). The initial stages are simply to remove the epoxy above the component, and the transition to 400-grit should be made when the component shadow appears in the epoxy. Any contact between the sample and abrasive papers coarser than 400-grit, will damage the component. Fine grinding follows, using 400 and 600 grit abrasive. Sectioning components down to 600 grit papers generally gives acceptable defect clarity and definition between the various materials used to construct the device. To improve the defect image and remove the smallest sectioning scratches, a final polishing step can be introduced. A 50μm grade diamond paste polish on a rotating cloth is adequate. For photographic purposes, dipping the sectioned surface in nitric or sulphur acid for five seconds improves the surface contrast further.

An example of a sectioned sandwich construction silicon diode is shown below (Figure 10-3). Sectioning through the X - Y plane clearly shows detail of the silicon die, sandwiched between the lead frames, and the die bonding solder fillets.

During sectioning, the analysis engineer will repeatedly observe the sample with an optical microscope to identify defects; for example, epoxy to lead frame delamination, encapsulation cracks or deformation of the die attach solder fillets. Figure 10-4 illustrates a symptom of excessive current, causing die overheating and solder fillet melt. Under pressure, the molten solder is forced down between the die and epoxy creating a short circuit. Alternatively, during solder cooling, it draws back into the fillet leaving a trace of carbonised epoxy and a consequent current leakage path. Solder fillet melt can often be identified during X-ray observation by the presence of solder or solder deposits in the image.

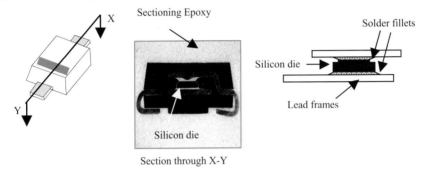

Figure 10-3. X – Y Sectioned sandwich construction silicone diode

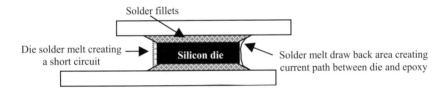

Figure 10-4 The effects of die attach solder melt producing short circuit or high leakage currents across the silicon die.

Electronic component sectioning is generally favoured for high magnification examination of passive components, such as capacitors, resistors and inductors. The process is also suitable for the internal examination of transistors, diodes and even integrated circuits, but because of the construction complexity of these devices, it is often necessary to completely or partially remove the component encapsulation material by a process known as package opening.

2.2. Encapsulated package opening

Package opening involves using aggressive acids to remove the epoxy packaging from around the component to allow examination of the silicon die, bond wires and lead frame. Figure 10-5 illustrates the traditional manual process of package opening.

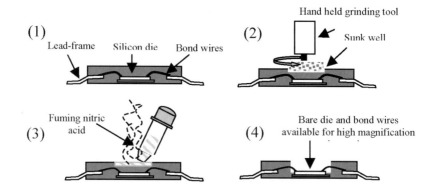

Figure 10-5. Traditional method of opening integrated circuit packages

A grinding tool is used to form a well in the epoxy molding material (10-5.2) directly above the die area. The well is then filled with fuming nitric or sulphuric acid (10-5.3), washed after a short interval and replenished several times until the area for examination is cleared of the encapsulation material (10-5.4). In addition to difficulties with health and safety regulations, the process is unreliable, as over-exposure to the acids can result in total destruction of the device die and bond wires. This is a particular problem for low profile high-density wire bonded integrated circuits. Automated package opening equipment is now available in which the de-capsulation process is confined to a sealed chamber and the component exposure time to the acid precisely controlled. Figure 10-6 shows a commercial automated package opener. Automated package opening equipment can de-capsulate a large integrated circuit in approximately two minutes, depending upon the epoxy volume and the acid strength.

Figure 10-6. Automated package opener NSC Corporation. Model PA103

Although not excessively expensive, package-opening equipment can be costly to run and the mechanical integrity of the system maybe affected by corrosive acids. Figure 10-7 illustrates the process block diagram showing acid and water flush cycle paths.

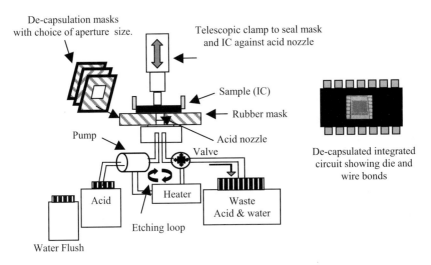

Figure 10-7. Automated acid dispensing and wash cycle, package opener

The system comprises a set of reusable masks offering a choice of aperture dimensions, a sample chamber with a pressure adjustable telescopic clamp and an acid heating and circulation system. The IC and rubber mask, with an appropriately sized aperture are positioned over the acid ejector. A small amount of acid is drawn into the etching loop and continually cycled through the heater, loop control valve and mask aperture. The acid passing into the mask aperture slowly dissolves a window in the component epoxy until the component die and bond wires are exposed. At the completion of the acid cycle, the equipment automatically dispenses the used acid and repeats the cycle with water to flush the system and the IC cavity.

Small discrete components can also be opened using automated package openers, in which a hollow glass bell containing the component is sealed against the acid nozzle (Figure 10-8). Before opening small components, it is advisable to solder the component to a miniature circuit board or solder all the terminations together by wire links, to provide support when the epoxy encapsulation is removed. Figure 10-8 also shows a de-capsulated miniature dual transistor where the component has been taken from the etching bell before all the encapsulation has been removed. The remaining epoxy supports the lead-frame, both die and bond wires, which are clearly visible.

3. UNDERSTANDING COMPONENT FAILURE MODES

Chapters 7, 8 and 9 have considered component construction, printed circuit board assembly and the tools of failure analysis. So far, this chapter has described component preparation for high magnification examination. The remainder of this chapter will present examples of component failure explaining the failure mode, diagnoses and suggest corrective actions to prevent re-occurrence.

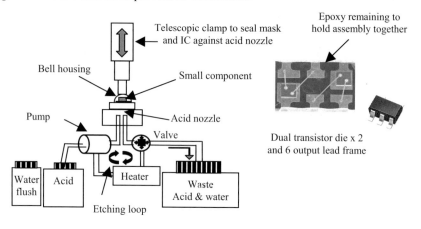

Figure 10-8. Automated package-opening system for miniature components and de-encapsulated dual transistor.

The prime objectives during the design, manufacturing and testing phases of electronic components, are to produce a device that meets electrical and mechanical specification and to ensure that no part of the manufacturing process compromises the reliability of the component. In a well-defined, controlled and closed component-manufacturing environment this is possible, but outside this environment, the component has to run the gauntlet of several potentially fatal situations during transit, storage, circuit board assembly and application testing. Some of these include electrostatic discharge (ESD), mechanical handling abuse, thermal and electrical overstress. Component users have confidence that components will survive these hostile environments because they have been manufactured and tested in accordance with internationally accepted standards that ensure a minimum standard of durability. Some examples of component failures induced by electrical, thermal and mechanical misuse and those propagated via the component manufacturing process are now presented.

3.1 Mechanical and thermal stress

Surface mount chip resistors, capacitors and inductors are particularly fragile in the sense that the ceramic substrate is relatively strong but exhibits almost zero flexibility (see Chapter 2). These components are sensitive to mechanical stress during PCB assembly, soldering (differential coefficients of expansion) and PCB distortion during handling and attachment to the final application.

Component cracking can be induced during PCB assembly, as a result of excessive placement travel and force, possibly accentuated by excess solder paste on the PCB solder pads. Residue, collecting on the base of the pick up nozzle, is also a major contributor to component placement fractures. Damage can break the component through its body or cause microscopic hairline cracks on the surfaces that penetrate through the device. For example, what appears to be a surface crack may travel through the component body and separate the plates of a multilayer chip capacitor or a chip resistor element. The component may function normally while the inner construction continues to make electrical contact but it fails subsequently during mechanical or thermal stress. A capacitor or resistor exhibiting failure due to internal cracking may disintegrate when de-soldering it from the PCB. This is a further reason for analysing components before they are removed from the PCB, as the appearance and position of the crack is often an indicator of the failure root cause. Figure 10-9 illustrates typical failure mechanisms introduced at the component placement operation. Component placement breakage may also be the result of poor set up of the pickup nozzle travel or marginal upward bow in the PCB from the flat plane.

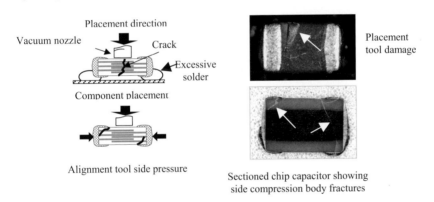

Figure 10-9. Multilayer chip capacitor failure due to mechanical damage

Chip resistor components, reported as being out of tolerance (high resistance), are usually the result of microscopic cracks travelling through the glass and resistor element. Examination of the surface layer will reveal the defects. Figure 10-10 demonstrates how poor handling or location of fixture points can mechanically stress components. It is preferable for components to be orientated so as to minimise stress induced by board flexing. Component A, in Figure 10-10, experiences a small bending stress, while component B, which has its long edge on the bend radius, can potentially fail due to cracking or termination tears where the inner terminations of the component separate. Components B and C are placed too close to the PCB fixing point. The PCB will be distorted by the fixing bolt or clamp, with stress permanently transmitted to the component solder terminations. Termination tear can also be induced when solder joints solidify while the PCB remains warped after the pre-heat and soldering zones. During cool down, the warp will relax, the PCB straightens and tension is applied to the solder joints. This stress is minimised by complying with industry standard soldering profiles (see Chapter 8, Sections 7 and 8) designed to reduce the dimensional differential between all component parts and materials during the soldering process. PCBs are generally clamped during their travel through the soldering process to minimise board warpage.

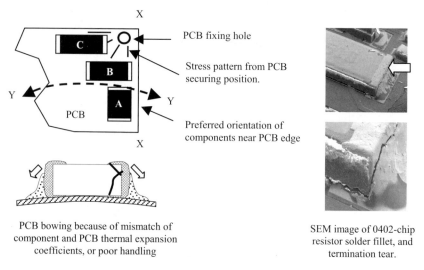

PCB fixing hole

Stress pattern from PCB securing position.

Preferred orientation of components near PCB edge

PCB bowing because of mismatch of component and PCB thermal expansion coefficients, or poor handling

SEM image of 0402-chip resistor solder fillet, and termination tear.

Figure 10-10. Component mechanical damaged induced via PCB fixings

The effects of termination tear can be serious, as the electrical continuity of the torn soldered termination may be sufficient to pass the electrical final test. Failures occur later when the inner tear oxidises and becomes open circuit. For this reason, manufacturers of high reliability circuit board assemblies, subject them to simulated drop tests and vibrations.

3.2 Electrical overload

External damage to a component may be the result of a high-energy transient over-load that generates heat within the component. Figure 10-11 shows encapsulation damage in a two terminal sandwich construction diode. The crack looks similar to termination tear, but X–ray examination indicates that the die bonding solder had melted and that a crack ran from the die area out to the point where the termination exited from the component package molding.

Sandwich construction diode assembly produces a package that does not exhibit the weakness associated with utilising gold bond wires to connect the die connection points (die bonds) to the respective lead frame finger. During overload, bond wires act as a quick fuse, but in the sandwich construction diode, the package absorbs transient power overloads. However, if the overload persists, die and lead frame will overheat, expand

and fracture the package as shown in Figure 10-11. Figure 10-12, shows failure of a wire-bonded transistor due to transient overcurrent.

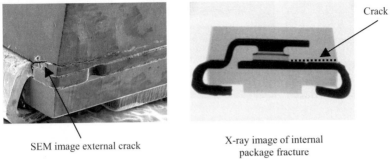

SEM image external crack

X-ray image of internal package fracture

Figure 10-11. Component external damage and X-ray image of silicon die and lead-frame

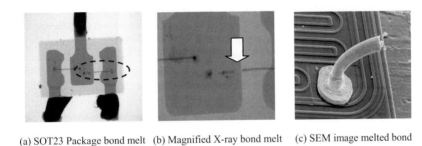

(a) SOT23 Package bond melt (b) Magnified X-ray bond melt (c) SEM image melted bond

Figure 10-12. X-ray images of a SOT 23 transistor package with open circuit base to emitter bond wire connection.

In figure 10-12(b), the bond wire appears to be open circuit from a small gap in the X-ray image. However when the transistor was opened and viewed under the SEM, it was evident that almost all of the gold wire had evaporated during the high temperature transient. The image is not sharp because the track was actually carbonised epoxy in the area where the bond wire was encapsulated before evaporating. When viewing the internal construction of any product, it is important to remember that evaporated metallic parts can appear as near-normal, particularly in integrated circuits where bond wire melt is a common failure mode.

The effects of excessive overvoltage transients on components may vary considerably. Contrary to popular opinion, chip resistors can have their characteristics modified by overvoltage. During these conditions, the electrical energy breaks down across the laser cut, and partially reconnects the previously clean laser channel. This effectively reduces the resistor value as measured across its terminations. Figure 10-13 shows a standard 0805 chip resistor (104 = 100K ohms) that was returned for failure analysis because it measured 88.3K ohms across its terminations. After examining the resistor for termination tear and mechanical damage, it was sectioned to remove the top and middle glass layers to enable examination of the laser trim channel. (Figure 10-13(b)).

 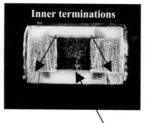 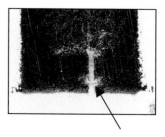

(a) 0805 chip resistor.   (b) Glass removed to view laser trim.   (c) Magnified laser trim melt.

Figure 10-13. Standard 0805 chip resistor sectioned to remove the top and inner glass layers, to view laser trim condition.

At higher magnification (Figure 10-13c), the laser track is shown to be short circuited by melted glass and conductive material from the resistor, ruthenium oxide, element. This partially reconnected the laser track and reduced the resistance value measured across the component terminations. This is a well-known failure mechanism caused by short duration high energy transient overvoltages. The industry standard, 0805-chip resistor is specified to withstand a maximum working voltage of 150 volts. A voltage between 1000 and 2000 volts, possibly through a lightening strike, would induce the failure shown in Figure 10-13. If the overload voltage and therefore current, was maintained for several milliseconds, very high internal current and temperature would be generated, causing the resistor element and glass protection layers to melt and outgas through the top surface of the component. Figure 10-14 shows evidence that this is what has occurred with the internal heat and subsequent pressure released via the top surface blowhole.

 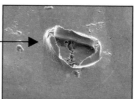

Figure 10-14. Surface mount chip resistor showing external damage caused by internal power overload melt.

3.3 Electrostatic Discharge (ESD)

Electrostatic discharge voltage is a problem that continually plagues reliability targets, particularly as component dimensions and silicon geometries continue to shrink. Historically, complementary metal oxide semiconductors (CMOS) were considered sensitive to ESD because of the small dimensions (gate widths) and very thin junction layers. Bipolar devices utilise a technology which is considered more robust and almost immune to ESD damage. The drive for miniaturisation has caused MOS technology parts to be more sensitive to static discharge, and the reduced bipolar geometries have brought these devices into the category of being ESD sensitive.

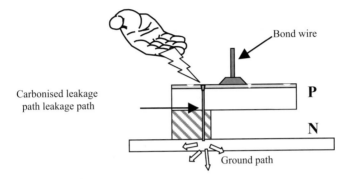

Figure 10-15. Fundamental illustration ESD path through a PN junction, leaving a leakage or short circuit path across the junction

The fundamental ESD failure mode is demonstrated in Figure 10-15, where an electrically charged human is discharged into an electronic component PN junction. The static electricity will discharge to ground where it finds a point that cannot withstand the overvoltage potential. If this is through the semiconductor junction, the discharge path will intensely heat a passage through the junction leaving a track of carbonised materials and, subsequently, an unwanted leakage path across the junction. Where the discharge voltage potential is not great enough to break down the junction, or a low resistance path is found around the device, the static charge will dissipate harmlessly. Semiconductor manufacturers design ESD protection networks into their devices to meet the ESD standards organisations requirements. JEDEC, ESD Association and BS EN 61340-5-1 have jointly issued standards with which component manufacturers must comply, and categorise their components into degrees of static sensitivity. Generally, components will withstand 2000 volts whereas some such as CD laser diode detectors will be as low as 20 volts. In this case, handling precautions must be very well defined and regulated within ESD safe handling areas. The standards organisations specify static discharge capability under three environmental situations. Human Body Model (HBM), Machine Model (MM) and Charged Device Model (CDM). The component manufacturer will specify which condition the ESD sensitivity is measured against.

The human model test (HBM) simulates the natural resistance of a human charged to a voltage level, and the discharge rate into the component pin being restricted by the human resistance (200 ohms). The machine model (MM) is more aggressive and simulates a fixture or fitting electrically charged and brought into contact with the component. The resistance is assumed to be zero, and therefore the static discharge will be very rapid and more dangerous to the integrity of the component. The charged device model (CDM) simulates the actual component being charged and brought in contact with a low impedance ground. This is similar to the machine model where the discharge time is very fast.

Manufacturers product reliability data give maximum ESD voltages for assemblies or components, and are specified as follows, HBM 2Kvolts, MM 200 volts, and CDM 1000 volts. Either preventing the charging process or providing a safe discharge path may control ESD. Materials to prevent charging are called antistatic, and those used to safely discharge voltages are static dissipative.

The modern PCB assembly area is a totally contained ESD protected area where strict procedures must be followed. The floor and all materials in the area are antistatic, and all component packaging is static dissipative. Personnel clothing will comply with ESD standards and all food containers (plastic cups) banned from the safe handling area, as these are perfect static generators. Warehouses are also created to provide a safe static storage environment for components via personnel training for the handling of electronic components and the need to wear wrist and heel straps whenever components reels / boxes have to be handled. Components must never be removed from their individual tubes, reels or bulk boxes unless they are taken to the specified safe handling area and all other procedures followed.

Despite great efforts by component manufacturers and PCB assemblers, components are still damaged by ESD but fortunately the percentage is extremely low. ESD failures are more commonly induced in the field but although the occurrence is minimal, the effect on product reliability is significant in terms of the time taken to determine the root cause and implement and monitor a corrective action. An example of the major impact that ESD damage can have on product reliability and the corrective action to prevent reoccurrence is presented below.

A customer, purchasing a standard surface mount transistor for design and build into a home security system, experienced random field failures diagnosed as increased leakage between collector and emitter of a bipolar transistor (small collector current when the transistor should be switched off). When the failed transistors were returned to the component manufacturer for analysis, it was found that the parts were out of specification. Production, lot code traceability searches were implemented and 'Keep' samples of these lot codes retested to confirm that the parts contained the correct die, and that they fully complied with the electrical specification. Similar lot codes shipped to other customers were checked and found to be satisfactory. The failed parts were again subjected to X-ray and electrical function test where high leakage was confirmed. De-capsulation and high magnification examination identified microscopic craters on the surface of the transistor die. Further specialist sectioning revealed that the microscopic craters were the top ends of carbonised tracks (Figure 10-15) through the silicon i.e. ESD damage. Figure 10-16 shows SEM images of the ESD craters. Image (a) shows emitter and base bond wires and the silicon die soldered to the lead frame, (c) shows increased magnification of the emitter bond wire ball attachment to the silicon pad with images (b) and (d) showing further magnification of areas marked x, and y (Figure10.16, b and d). ESD punch-through craters are clearly identified.

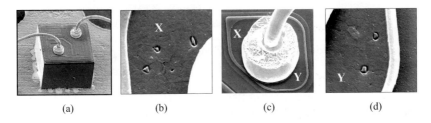

(a)   (b)   (c)   (d)

Figure 10-16. Sectioned bipolar transistor showing SEM images of ESD crater damage.

Further investigation revealed that in the application circuit, the collector of this transistor was connected directly to the outside world via the PCB connector, with no overvoltage protection clamps. It was concluded that electrostatic discharge was being applied to the transistor via the connector pin during handling and installation. ESD was inflicting damage when the static voltage level exceeded the capability of the device. All PCBs in the field and new production boards were modified to incorporate overvoltage suppression diodes on the connector pin connected to the transistor, in order to surge suppress any voltage levels on the pin greater than 10 volts. This corrective action completely cured the problem and increased the assembly handling durability and reliability.

A second example of ESD destruction, involves the failure of a CMOS integrated circuit in an environment where the PCB assembler, component manufacture and supply chain ESD handling procedures were exceptionally good. For several months, the PCB assembler passed a high volume of this IC through the production line with zero defects. Suddenly, they experienced a high percentage of failures at the automated electrical final test station where the IC was rejected for high leakage on several input pins. Despite the PCB assembler's excellent ESD handling procedures, the component manufacturer's analysis attributed the failures to ESD damage.

Figure 10-17 shows the IC encapsulation removed, and the die area and bond wires available for examination. The SEM identified an ESD pinhole, which created a low resistance path between the topside aluminium tracking layers and internal silicon. This was detected as higher than expected leakage current during the PCB assembly functional test at a specific PCB test point. Fortunately, the assembler had excellent traceability for all components placed, and also traceability through to machine and operators. The primary cause of failure was identified as one operator being unofficially allowed to wear none ESD compliant overgarments because of an allergy to antistatic clothing materials. This was compounded by poor discipline in wearing a static discharge wrist strap during handling of components and PCB assemblies. In this case, the PCB assembler had first class ESD protection facilities, but did not implement an effective auditing system for total compliance by the operators.

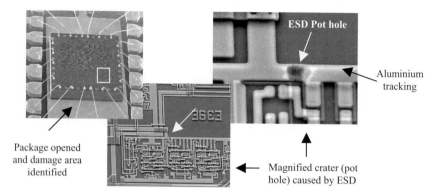

Figure 10-17. Scanning Electron Microscope images of integrated circuit metallisation and internal silicon damage caused by electrostatic discharge

Analysis of overvoltage failures can be difficult where the failure mode does not consist of a heavily burnt out and destroyed area, which is a clear indication of over heating due to power overload. Sometimes, the failure analyst encounters damage by overvoltage, which may have been caused, for example, by applying 12 volts to a 5-volt termination, as opposed to ESD damage. This is a credible failure mode when a newly assembled PCB has a solder short between two or more terminations. Figure 10-18 illustrates SEM images of an opened integrated circuit where damage was detected on two metallised tracks (open circuit).

Metalisation failure area.   Carbonised encapsulation epoxy

Figure 10-18. SEM images of damaged metallisation layer of an integrated circuit caused by over current.

Electrostatic discharge pulse durations are very short but long enough to intensely heat buried layers of the semiconductor device. Generally, the damage is highly localised as shown in figures 10.16 and 10.17. Where the damage is caused by an accidental over current from a source that can sustain longer duration pulses and energy, additional carbonisation of the epoxy in close proximity to the damaged area will occur. During package opening, this carbonised epoxy is more resistant to fuming nitric acid and remains intact at the end of the package opening cycle. The presence of carbonised epoxy deposits indicates that the failure mode generated intense heat for several milliseconds, and was very probably overvoltage rather than ESD.

Overcurrent failures can cause serious damage to the component construction and leave little doubt to the cause of failure. Exceptions occur where, for example, wire bonds are disconnected but because no other damage has occurred, other sources of failure, such as mechanical stress must be considered. Figure 10-19 shows an X-ray image of a surface mount transistor with an open circuit bond wire.

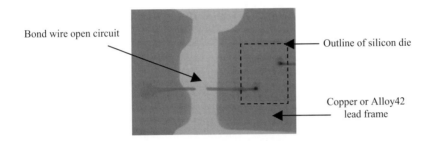

Figure 10-19. X-ray image of open circuit gold bond wire in a SOT23 transistor. Did it melt or did it fracture with mechanical or thermal shock?

Experience suggests that the bond wire loop melted due to current and heat overload. During the heat transient, the lead frame and silicon die on either side of the loop conducted heat away from the wire adjacent to it, but allowed the centre of the wire loop to overheat and melt. Additionally, when the package was opened and viewed, a ball of gold was apparent at each end of the wire where the loop has melted. Alternatively, where a component manufacturer has moved to an aluminium bond wire, the aluminium wire melts with a zigzag profile whereas the gold wire remains smooth (Figure 10-20), both phenomena indicating bond wire melt and not fracture via mechanical or thermal cycling stress.

Gold bond wire-melt characteristic.   Aluminium bond wire-melt characteristic

Figure 10-20. Illustration of wire bond melt characteristics between gold and aluminium bond wires.

If the bond is broken immediately adjacent to the lead frame or silicon die bond connection pad, failure is probably due to mechanical or thermal cycling stress. Bond wire shearing accompanied by no other damage, such as delamination or epoxy cracking, is probably due to mechanical vibration or deceleration shear forces as in a drop test. Contamination of the die bond pad is also a possible cause of wire bond failure, where the ball disconnects from the die pad. This type of failure is traditionally known to as the Purple Plague, and is particularly troublesome, as the device will operate satisfactorily at ambient temperatures but fail in open circuit as the package and die expand with increased temperature (Figure 10-21).

Contamination on the die bond pad will allow the wire ball bond to separate from the die bond pad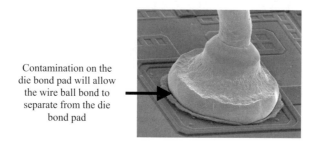

Figure 10-21. Wire bond ball joint and die bond pad connection area where contamination can affect the long-term reliability of the joint.

Purple plague is a serious issue in a semiconductor-manufacturing environment, and would certainly result in a very high profile program to check all materials, process parameters and procedures throughout the component assembly line. Any concerns that the failure mode was due to a process failure would result in component recalls from the customer base and screening of stock in the distribution network. There is really no recovery from the component point of view. The only course of action would be to scrap all similar production lot codes from that production facility, and possibly lot codes following and preceding the suspect batch.

3.4 Effects of contamination on electronic assembly reliability

Mismatch of material coefficients of expansion between PCB and components during soldering or during storage, in conditions outside their specified moisture sensitivity level (MSL), can affect package integrity via long term susceptibility to moisture ingression and weakened solder joints. Figure 10-22 shows magnified termination images of an 84-pin quad flat pack (QFP) semiconductor device that had been selected for use in a low cost application utilising a PCB with poor dimensional stability when cycled across a product operating temperature range of 0 to 85°C. Several failure modes were identified but all were induced by the severe alternating stresses exerted on the component leads as the PCB expanded and contracted. Figure 10-22 shows how the epoxy resin has separated around the component lead-frame, as the lead and component body moved by dissimilar amounts during thermal cycling.

Internal X-ray, acoustic micro imaging and package opening revealed sheared wire bonds, cracked silicon die and delamination between the lead-frame and epoxy. There were also signs of corrosion, induced by moisture ingression via the epoxy to lead-frame gaps. Acoustic micro imaging confirmed that several devices suffered serious delamination between the semiconductor die and the encapsulation epoxy. Figure 10-23 shows the scanning laser acoustic micro image (SLAM) in which the dark central area identifies the delamination void while the remainder is white.

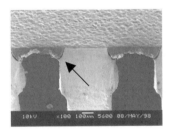 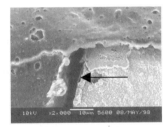

Figure 10-22. 84 pin QFP plastic package showing separation of the metal termination from the molding material.

An audit conducted by the component manufacturer demonstrated that the operating procedures in this customer's facility were basic and uncontrolled. The storage environment for components averaged around 27°C and 65 percent relative humidity. Components were held in their storage area for up to three weeks. The 84pin QFP device has a moisture sensitivity level (MSL) of level 2. That defines a maximum storage condition and time, as specified by IPC/JEDEC J-STD-020A, of <30°C/60%RH for 168 hours (7 Days). During storage, the QFP package was absorbing excessive moisture, which boiled and turned to steam during the soldering process, generating pressure and damage within the device. The company dramatically reduced the failure rate by improved storage conditions and including two additional fixing points on the PCB to reduce board flexing and warping across the QFP termination area.

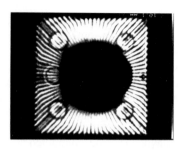

Figure 10-23. Scanning laser micro image (SLAM) of an 84 pin QFP package with serious internal delamination void.

3.5 Contamination and environmental effects on electronic assembly reliability

Electronic circuit boards continue to reduce in size, in parallel with an increasing component density per unit area. The associated reduction in the pitch between component terminations has focused attention on the need to ensure that PCB materials, solder resists and cleaning chemicals do not leave residues that cause long term electrochemical failures due to leakage paths under the solder resist and between component terminations. Measuring the surface insulation resistance (SIR) of a printed circuit board indicates the probability of electrochemical failure at a later date, and hence it is a quality control tool to ensure long-term reliability more than a failure analysis tool.

Three elements are required to provide the conditions for electrochemical failure; humidity, ionic contamination and electrical bias. Where these conditions exist, metal migration occurs under the solder resist, creating electrically conductive dendrites between component terminations.

Figure 10-24. SEM image of dendrite growth due to migration.

Figure 10-24 shows such a PCB dendrite resulting from contamination and long-term metal migration under the PCB solder resist.

4. CONCLUSIONS

The electronics assembly industry is built on long-term experience and know-how, supported by international standards to ensure a minimum level of quality from PCB, raw material and component suppliers. Despite these controls and procedural disciplines, component failures do occur during product assembly or in use. Zero defects and product reliability is the goal of all manufacturing companies within product design and build. This chapter has emphasised the importance of accurate monitoring during production and the care required in the preparation of specimens for failure analysis. With a series of practical examples, it has demonstrated their cause and significance, plus strategies to avoid reoccurrence. Table 10.1 lists common component defects associated with silicon die, die-attach, wire-bonds and plastic/lead-frame integrity that affect product reliability.

Die	Die-Attach	Wire bonds	Plastic & Lead frame
Electrical overstress	Thermal Stress	Electrical overstress	Moulding stresses
Electromigration	Delamination	Tensile fracture	Package warpage
ESD damage	Voiding	Mechanical fatigue	Cracking
Cracking	Shorts to Active Area	Power / thermal stress	Delamination
Corrosion	Outgassing/Moisture	Intermetallic – Purp Plague	Moisture Absorption
Diffusion/Oxide problems	Dendrite Formation	Kirkendahl Voiding	Ionics
	Fatigue		

Table 10-1. Typical component failures associated with build and materials

RECOMMENDED READING

1 P L Martin, Electronic Failure Analysis Handbook, McGraw Hill, 1999
2 SMART Group web site *www.smartgroup.org*

CHAPTER 11

MAKING THE TRANSITION TO A RELIABLE LEAD-FREE SOLDER PROCESS

1   INTRODUCTION

It is surprising that about 30 percent of the lead associated with a typical joint does not originate from the solder. So, lead-free soldering means that the component external electrodes, the PCB surface finish, as well as the bulk solder, do not contain lead. A product will be considered lead-free when any of the individual materials used to construct it, contains less than 0.1 percent lead. The transition to lead-free also requires that the component and PCB have the capability to withstand the increased temperatures associated with lead-free soldering.

The elimination of lead presents several challenges to identify a replacement alloy that is close to a drop in replacement, with respect to melting point, processability and mechanical durability (reliability), **and** utilises existing PCB assembly equipment. Chapter 1 described he background to lead-free solders and proposed potential replacements. The present chapter recommends the technology and logistical route to make the transition to a lead-free soldering process. It also gives examples of reduced joint reliability where the lead-free alloy has become contaminated from lead bearing PCB finishes and component electrodes. Within the environmental legislation to remove lead from the electronics industry, there are exemptions where no viable lead-free solution is yet available. These include glass, high temperature solders and mission critical equipment, comprising servers, certain key storage and life critical medical equipment. The exemptions will remain in force until 2010, when the long-term reliability of lead-free solders will be reviewed.

2   RELIABILITY CONCERNS DURING LEAD-FREE SOLDERING

2.1 Contamination of lead-free alloys

Tin and lead have melting points at 323 and 232°C respectively, and their alloys have lower melting points, with a minimum of 183°C occurring at the eutectic composition (63Sn-37Pb). The phase diagram for tin and lead was presented in Chapter 1 (Figure 1.5). It is notable how the melting point of tin is dramatically reduced by a small addition of lead.

Lead-free soldering is not a new concept, as gold and nickel palladium (Ni-Pd), have been used for component termination plating for several years. PCB finishes have typically utilised a hot air solder level (HASL) process to lay a thin layer of Sn-Pb solder onto PCB copper tracks. However, as this process produces a relatively uneven surface finish, it does not comply with the planarity requirements (all terminations simultaneously touching the track) of large fine pitch components.

Alternative smooth PCB finishes, such as nickel gold, immersion tin, organic solderability preservative (OSP) and more recently, immersion or electroplated silver have been in use. These finishes are coincidentally lead-free. OSP is a temporary organic finish, which is applied directly to the PCB copper board to prevent oxidation of the copper. It vaporises during the soldering process as the component is soldered directly onto the copper track. Lead-free HASL is also now available but is still eliminated from applications using fine pitch devices because of the irregular surface finish that is produced.

Long-term experience (1), during the use of lead-free PCB and component finishes, has proven joint reliability when used with lead based solders. Figure 11-1 gives percentages of solder and plating volumes for a typical solder joint where the base solder dominates the volumes.

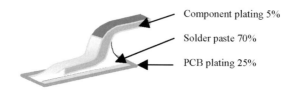

Figure 11-1. Typical surface mount solder joint and material volumes.

Lead from the solder will defuse into the 10 to 20μm thick component plating essentially, forming a leaded-joint with proven reliability. The solder will also form an intermetallic layer with the PCB finish.

Introducing leaded-component terminations into a lead-free solder process is an important consideration, as it is very probable that they will be used in a lead-free process during the transition to a totally lead-free system. The lead in this situation is effectively a contaminant of the lead-free solder. During cooling, the final liquid to solidify is rich in lead. Figure 11-2, demonstrates how the lead from the component plating does not diffuse into the bulk solder but collects and forms a lead-rich area. This region exhibits inferior thermal cycling fatigue characteristics than the original lead-free alloy. It has been reported (2) that only one percent of lead contamination can reduce the thermal fatigue of a lead-free solder joint by greater than 50 percent. Also, the resultant alloy Sn-Ag-Cu and Pb, has a lower melting point than the base solder, resulting in the possibility of de-wetting during subsequent soldering operations.

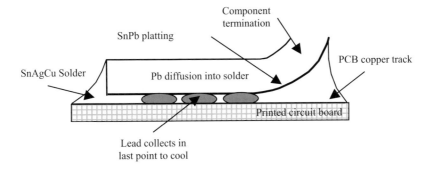

Figure 11-2. Lead contamination during termination cooling.

Lead contamination of a lead-free solder bath can also result from the use of leaded terminations in a lead-free process. As indicated earlier, the melting point of tin is readily depressed by small additions of lead. A more serious combination, in both surface mount and wave soldering, is lead contamination in conjunction with bismuth, either deliberately in the bulk solders or again, as contamination from component or PCB finishes. Pb and Bi can produce an alloy with a melting point of less than 100°C, producing a joint with a low melting point and a high probability of re-flowing under normal product operating temperatures.

Further concerns about using leaded component terminations in a lead-free process, involve the components resistance to the increased soldering temperatures necessary for lead-free alloys. These maybe 30°C higher, and impair the component's long-term reliability if they have not been qualified to such increased temperatures. All lead-free components will be requalified by the component manufacturer to withstand soldering temperatures (soldering durability) of upto 260°C, whereas previously, they were qualified to typically 250°C. This upper ceiling on the operating temperature level is imposed by the mechanical behaviour of the polymeric materials around the components and in the board (Chapter 2).

The recommended transition route to a lead-free process is to introduce lead-free component terminations into existing assembly processes and to continue the use of leaded bulk or paste solders. Reflow / flow profiles should be maintained, and the change to a lead-free solder should be made only when all components and PCB attachments are lead-free. At this time, the process parameters can be changed to be inline with recommended lead-free soldering conditions.

2.2 Fillet lifting

Solder fillet lifting is a problem associated with through hole component technology and, as the name implies, is associated with the solder fillet lifting away from the PCB land at each through hole via on the PCB surface. Figure 11-3 illustrates the process, and also shows the SEM image of a typical through-hole termination exhibiting the phenomenon. From Chapter 1, eutectic Sn-Pb solder has a well-defined melting point and, unlike other compositions in the system, solidifies rapidly at a single temperature. A disadvantage of lead-free solders is their higher surface tension when in the liquid state and wider temperature range between being fully solid and totally liquid. This range is known as the mushy or pasty range.

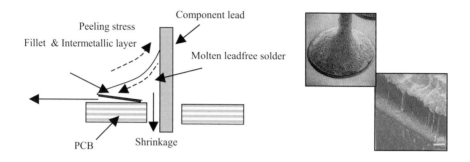

Figure 11-3. Mechanism of fillet lifting and SEM image of a lead-free through-hole joint.

Fillet lifting occurs during the cool down process in through-hole, wave soldered, lead-free joints. Due to the wider solid to liquid range, lead-free solder joints tend to solidify at different rates depending on the heat sinking capability of the component and PCB. Immediately after leaving the solder wave, the component lead will draw heat away from the PCB through-hole barrel, encouraging the solder in the barrel to solidify and shrink. During the cooling phase and while the solder adjacent to the PCB land is still molten, the cooling and shrinking solder, from the component lead outwards, pulls the solder away from the PCB land along the intermetallic layer. The degree of fillet lifting increases with board thickness and the extent of the pasty range of the solder. Bismuth containing solders exhibit almost 100 percent fillet lifting on a typical PCB. Joint reliability is of primary concern where fillet lifting occurs. Although fillet lifting will not be acceptable in many applications, current reliability testing has shown that the pull strength of the lead-free through-hole solder joint exceeds that of leaded solder despite the fillet lifting.

2.3 Tin whiskers

The introduction of lead-free solders has precipitated the use of high-tin content, component termination plating and PCB finishes. Surface mount, passive component terminations are being replaced with an electroplated layer of pure tin. Tin is low cost and exhibits excellent shelf life and solderability. It is, of course, non-toxic. The concern of many PCB manufacturing companies is the tendency for microscopic spontaneous metallic filaments, known 'tin whiskers', to grow on tin. In high-density PCB assemblies where fine pitch component terminations are used, tin whiskers may cause transient or permanent short circuits in low voltage applications when insufficient current is available to melt the whiskers, as they connect between circuit points. Figure 11-4, illustrates the whisker phenomena and shows an SEM image of whiskers observed after 500 temperature excursions between –55°C and 125°C, at 30 minutes per cycle.

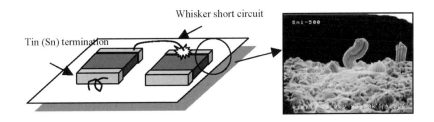

Figure 11-4. Illustration of whisker growth on tin termination, and SEM image of actual whisker, 20μm in length.

Whiskers can grow to several millimeters in length and up to 6μm in diameter (a human hair is typically 50μm in diameter). Growth rate is indeterminate; either occurring almost immediately after plating, or several years later. Observations have also concluded that growth continues from the base of the whisker and not from the tip, confirming that growth is associated with the characteristics of the tin plating.

Whiskers are not unique to tin and have been observed on aluminium, iron, nickel, lead and palladium. Their occurrence is more prevalent on materials that are under tension, as is the case with fine-grained, electroplated tin. Tin whisker growth is further promoted by the presence of additional stress such as near-surface scratches. Tin, plated onto brass produces maximum plating tension and, hence, a high probability of whisker growth. Pure tin-plating onto nickel, as used in resistor and capacitor component terminations, is noted as producing a lower surface tension. Whisker growth is not fully understood, but observations confirm that the spontaneous growth is more predictable at temperatures in the region of 50°C and humidity level of 50% RH. (3).

Fortunately, for the majority of the electronics industry, alloying pure tin plating with the base solder during the PCB soldering process reduces the occurrence of whisker growth to a probability level currently experienced with tin-lead plating. Heating and subsequent slow cooling of the tin plating also relaxes the plating tension, further reducing the probability of whisker growth. In applications using low temperature flexible boards, the potential for whisker growth will remain where component placement and attachment utilises silver-loaded conductive adhesives to make the electrical connection between the component tin termination and the PCB pad. In this process, no alloying of the tin occurs and the adhesive curing temperature of 120 – 150°C is too low to relax the component plating stress. The study of whisker growth has been accelerated with the increasing use of high tin content solders and funding of further research by NEMI (USA), ITRI (UK) and JEITA (Japan) organisations to provide a reliability test and measuring method for whisker growth. The report is scheduled for publication in July 2004.

2.4. Erosion of solder bath equipment.

The erosion of the solder bath and fixtures by high-tin solders is a recognised problem, but it will be accentuated by the increasing use of lead-free solders in wave soldering processes. The erosion ranges from slight to catastrophic, where solder bath damage has resulted in molten solder being dispensed onto the floor constituting a serious health and safety problem. The erosion rate is linked to operating temperature, movement rate and tin content of the solder. Typical erosion rates of Sn-37Pb at temperatures below 250°C are light, but above 350°C it is high. For typical lead-free alloys, contamination around 96% tin, corrosion rates are excessive above 350°C, but are unknown below this temperature. Experimentation by TWI Ltd. has shown that a 316 stainless steel, 6mm solder bath impellor shaft rotating at 1500rpm, was reduced to 5.66mm with significant random pitting after 30 hours at 400°C

In addition to the safety issues associated with bath erosion, contamination of the bulk solder by iron and other metals (aluminium, copper and brass) changes the appearance of the solder and increases its melting point promoting solder bridging and other defects.

Solder bath manufacturers are addressing the erosion problem by increasing the quality of the stainless steel used for the bath and coating moving parts with ceramic or hard enamel surfaces. Work is still ongoing worldwide to fully address this issue.

3   CHANGING TO A LEAD-FREE SOLDERING PROCESS

3.1 Process temperature profiles

A typical electronic assembly can experience a temperature of 150°C during 'burn-in' final test and during transient component fault conditions, where the die temperature can exceed 150°C. At the other end of the scale, the component manufacturer attaches the silicon die to the component lead frame using high-lead solders (>85% Pb) with a melting point of 300°C or above. The melting point of the solder used in the PCB assembly process must have a melting point mid way between these two temperatures to ensure that the component joints do not reflow during PCB fault conditions and that the die attach solder does not reflow during board assembly. Figure 11-5 illustrates the three combined soldering temperature ranges for PCB operation, soldering process and die attach.

In a typical process, a superheat zone of 32°C is provided between the melting point of Sn-Pb solder (183°C) and the minimum reflow soldering temperature of 215°C. This ensures that the solder is sufficiently hot to allow the formation of reliable solder fillets, and prevents solder bridging between component terminations. There also exists a total process window of 67°C between the 183°C melting point and the industry standard maximum wave soldering temperature of 250°C

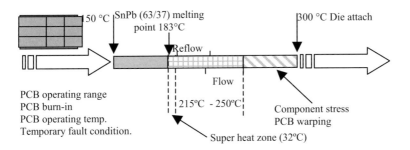

Figure 11-5. PCB tin-lead soldering process temperature range.

The immediate affect of changing to a lead-free solder, for example, Sn-Ag-Cu with a melting point of 217°C, is to reduce the superheat zone which ensures that the solder is sufficiently hot, and reduce the process window to 33°C. Lead-free solders exhibit greater surface tension than leaded solders. Therefore, the requirement for an adequate superheat zone is essential to ensure termination wetting and solder flow. The solution is to increase the maximum process temperature to 260°C and to raise the minimum temperature for reflow soldering to 235°C. This provides a superheat zone of 18°C and a total process window of 43°C. The increase in soldering temperature to 260°C is now internationally accepted by standards organisations, and requires manufacturers of bare PCB and component to re-qualify their products to this higher temperature. Figure 11-6 illustrates a typical revised lead-free temperature range for reflow and flow processes and the increased temperature durability requirement for lead-free components and the PCB. Ongoing trials, have shown that joint reliability can be achieved with a superheat zone of 12°C (4). This strategy requires very good process controls, as the process window is now particularly narrow.

Modification and recalibration of reflow and flow temperature profiles (see Chapter 8) is necessary when a process moves to lead-free soldering. All aspects of the PCB should be examined with respect to track width and land patterns, as the higher surface tension produces reduced flow in the molten state. Generally, changes to the PCB layout are minimal as long as the pre-heat and solder reflow zones are in line with the board layout and the solder manufacturer's recommendations. Although each PCB assembly process has individual characteristics, depending on board and component dimensions and layout, the surface mount reflow temperature profile illustrated in Figure 11-7 is typical of the current industry solution. This shows the modifications to surface mount lead-free soldering temperature profiles compared to those for lead-bearing solders.

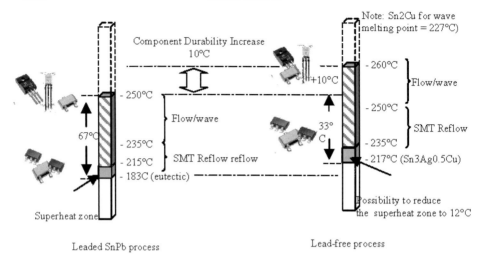

Figure 11-6. Comparison of leaded and lead-free soldering temperature range for flow and reflow processes

The characteristic transition from the previously flat preheat zone to a continuing slow increase, has been developed to ensure that all components and PCB have attained the same temperature when the PCB enters the higher temperature reflow zone. This eliminates any thermal gradients.

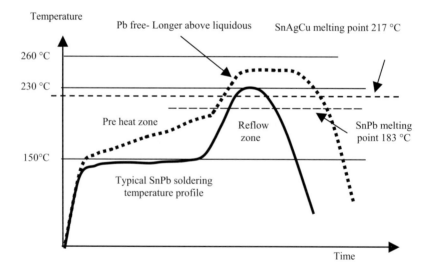

Figure 11-7. Comparison of leaded and lead-free soldering profiles

The increased length of the reflow zone is a compromise between minimizing reflow temperature and time to wet the component terminations because of the slower wetting time of lead-free solders. The cool down zone in a lead-free process is very important as the cooling rate determines the grain structure of the joint and its mechanical strength. The mechanical strength of a Sn-Pb joint was not so greatly affected by this rate.

4   CONCLUSIONS

Sufficient lead-free soldering trials have been conducted to demonstrate that lead-free soldering is viable and that the reliability associated with existing leaded joints can be matched. In some aspects, particularly with respect to thermal cycling durability, the new generation of lead-free alloys appears superior. This constitutes an important advantage with the continual reduction in joint dimensions and the requirement that the solder joint is the only way the component is attached to the PCB in surface mount applications.

To ensure maximum joint reliability during the transition to a total lead free process, it is highly recommended that the base solder remains leaded until all component terminations, PCB and fittings are lead-free. This route will avoid lead contamination problems of lead-free solders. Lead-free terminations in a leaded-process are also well proven and no change to joint reliability will be experienced. Several projects have changed to a lead-free solder while continuing to use leaded-components. They have achieved acceptable results but some have experienced joint failures, casting doubt on this strategy. Tin whiskers will remain a controversial subject for some time but component manufacturers generally consider this issue as one of component storage, and not a finished product problem. High reliability and military designs will certainly push for alternative component and PCB finishes to pure tin until the whisker issue has been resolved.

The duller grey and appearance of lead-free solder joints and fillet lifting in through-hole joints, although showing acceptable joint reliability, will be the subject of inspection retraining and automatic optical inspection (AOI) equipment recalibration.

REFERENCES:

1   Bob Willis, leadfreesoldering.com
2   K Seelig and D Suraski, Aim Corporation, A study of lead contamination in lead-free electronic assembly and its impact on reliability
3   The Growth of Tin Whiskers. International Tin Research Institute, Doc ref 734.
4   SMART Group web site. Lead-free section, www.smartgroup.org
.

PART 3

THE DESIGNER'S PERSPECTIVE

The strategy adopted in this Part is to explore the academic and research fields, more commonly linked with aerospace and power generation, with a view to evaluating the extent to which technology can be transferred to the interconnection situation. The principles of life prediction, particularly those involving crack propagation, under the time-dependent conditions relevant to service, are considered in some depth. Limited progress has been made, but a great deal more work is required.

Subsequently, the statistical treatment of empirical test data is considered. Again, the approach is elementary to cater for the diverse readership, although the practice of accelerated testing is subject to closer scrutiny from the physical metallurgical standpoint.

The concluding chapters outline the basics of finite element analysis and the challenges associated with time-dependent, unstable, materials, such as solders. The need to understand materials behaviour is equally strong in this field, as any other. A final chapter involves a detailed case study intended to illustrate the multifaceted and inter-related nature of the problem of structural integrity and reliability in electronics.

CHAPTER 12

INTRODUCTION TO LIFE PREDICTION

1 INTRODUCTION

The capability to efficiently design interconnections and to reliably predict their lifetimes during service are major goals of the electronics industry. Premature failure may have troublesome, expensive or disastrous consequences according to its nature. Having to rely upon empirical testing prior to market release, Industry is well removed from purely paper-based design. The trialling process generally involves thermal cycling of complete boards and their numerous component configurations, and for applications requiring low failure rates, may involve prolonged periods of testing. Time to market is increased with a consequent fall in competitiveness of the product. The present chapter outlines the general principles of life prediction, mainly as applied to bulk samples. In Chapter 18, case studies relating to actual joints will be presented.

2 GENERAL APPROACH TO MODELLING

Various modes of mechanical failure have been mentioned previously. While events, such as overload, vibration, bending and twisting do occur, the principal mechanism producing the majority of failures in service is thermomechanical fatigue (TMF). This process involves both cyclic (fatigue) and time-dependent (creep) behaviour, and employing the fundamentals presented earlier, life prediction under such conditions is now considered.

Life prediction models usually require information on the stress-strain behaviour of the sample or component under the anticipated operating conditions. For complex geometries, such as solder joints, experimental measurement of stress-strain behaviour is extremely difficult, and finite element analysis (FEA) is becoming the most efficient route for obtaining these relationships. This method is considered in greater detail in Chapter 16.

Modelling consists of the following steps:

- Constitutive equations are required to describe elastic and elastic-plastic deformation, which are time-independent. In addition, since room temperature is greater than one half of the homologous temperature for solders, time-dependent viscoelasticity and creep play an important role. The constitutive equations of both categories are derived from experimental data, usually obtained from bulk samples.
- The appropriate constitutive equation is used in a FEA program to calculate the stress-strain values throughout the component under the simulated service conditions. Details of commercial packages for this purpose are presented elsewhere.

- Finally, the FEA results are used in conjunction with expressions for lifetime to predict the life, and the results are compared to experimental thermal cycling data. The process, overall, is illustrated schematically in Figure 12.1

Examples of constitutive equations for time-independent, time-dependent and cyclic behaviour are considered in the following section.

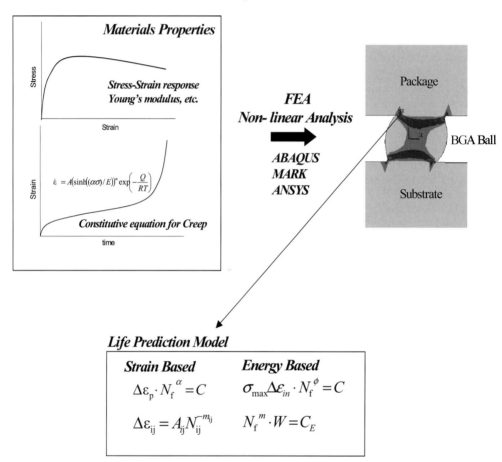

Figure 12-1 Fatigue life prediction steps using FEA for solder joints

3   EXAMPLES OF CONSTITUTIVE EQUATIONS

3.1 Time-independent Behaviour

As introduced in Chapter 2, time-independent behaviour includes elasticity and plasticity. In linear elasticity, stress and stain are related via the modulus (e.g. $\sigma = E\varepsilon$) whereas in the plastic region, the concept of work hardening was introduced. In this domain, equation 2.5 indicated that

$$\varepsilon = \sigma_y\Big/E + \left(\frac{\sigma - \sigma_y}{m}\right) \tag{12.1}$$

Where σ_y was the yield stress and m the work hardening coefficient. A special case, when m=0, was described as ideal elastic-plastic behaviour, since no work hardening occurred. Experimentally, the plastic region of the stress-strain curve for metals and alloys is generally curved rather than linear. This can be accommodated by a power law relationship (the Ramberg-Osgood expression) from equation 2.6 (Figure 12-2).

$$\varepsilon = \sigma_y\Big/E + \left(\frac{\sigma - \sigma_y}{a}\right)^b \tag{12.2}$$

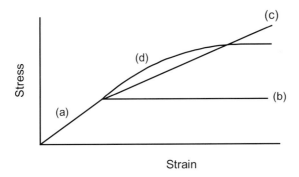

Figure 12-2. Constitutive stress-strain relationships (a) elastic; (b) perfectly plastic; (c) linear work hardening; (d) realistic work hardening

3.2 Time-dependent behaviour

Creep is a predominant deformation mechanism for solder joints, and for accurate modelling, FEA programs require reliable constitutive equations to calculate the stresses and strains in the joint. As indicated in Chapter 3, there are several types of creep deformation mechanisms – dislocation creep, diffusion creep and super-plastic deformation. Each mechanism has specific constitutive equations for the relationship between steady state (or minimum) strain rate, $\dot{\varepsilon}_m$, and applied stress, σ. For solder alloys in electronics applications, dislocation creep is usually the dominant mechanism. Diffusion creep is unlikely, since it generally occurs at very low strain rates and high temperatures. A generalised constitutive equation for creep can be written in the following form (1),

$$\dot{\varepsilon}_m = A\frac{DGb}{kT}\left(\frac{\sigma}{G}\right)^n\left(\frac{b}{d}\right)^p \tag{12.3}$$

where $\dot{\varepsilon}_m$ is the steady state or minimum strain rate, A is constant, G is the shear modulus, b is the Burgers vector, k is the Boltzmann constant, T is the temperature, d is the grain size, p is the grain size parameter, n is the stress exponent, and σ is the stress.

However, dislocation creep is independent of grain size, and the constitutive equation becomes (Dorn equation) (2),

$$\dot{\varepsilon}_m = A\frac{G\Omega}{kT}\left(\frac{\sigma}{G}\right)^n\frac{D_{eff}}{b^2} \tag{12.4}$$

where Ω is the atomic volume and D_{eff} is the effective diffusion coefficient. The effective diffusion coefficient is

$$D_{eff} = D_{SD} + b^2\rho D_p \tag{12.5}$$

where D_{SD} is the self diffusion coefficient, D_p is the pipe diffusion coefficient and ρ is the dislocation density. Since $\rho \propto (\sigma/G)^2$. Equation 12.5 is rewritten as:

$$D_{eff} = D_{SD} + \beta\left(\frac{\sigma}{G}\right)^2 D_p \tag{12.6}$$

where β is a parameter for dislocation distribution. So,

$$\dot{\varepsilon}_m = A\frac{G\Omega}{kT}\left(\frac{\sigma}{G}\right)^n\left\{D_{SD} + \beta\left(\frac{\sigma}{G}\right)^2 D_p\right\}\frac{1}{b^2} \tag{12.7}$$

In high temperature or low stress regions, $D_{SD} >> \beta(\sigma/G)^2 D_p$. So,

$$\dot{\varepsilon}_m = A \frac{G\Omega}{kT} \left(\frac{\sigma}{G}\right)^n \frac{D_{SD}}{b^2} \qquad (12.8)$$

In low temperature, or high stress regions, $D_{SD} \ll \beta(\sigma/G)^2 D_p$. So,

$$\dot{\varepsilon}_m = A \frac{G\Omega\beta}{kT} \left(\frac{\sigma}{G}\right)^{n+2} \frac{D_p}{b^2} \qquad (12.9)$$

If the condition (as-cast, cold worked, or heat-treated) of a material is chosen, G, b, Ω and β are fixed and the dislocation distribution should be constant. Therefore, these parameters are amalgamated into a constant, A. Furthermore, $D = D_0 \cdot \exp(-Q/RT)$, so Equations 12.8 and 12.9 can be rewritten into a more simple form known as *power law creep*.

For high temperatures or low stresses,

$$\dot{\varepsilon}_m = a\sigma^n \exp\left(-\frac{Q_{SD}}{RT}\right) \qquad (12.10)$$

Alternatively, for low temperature or high stress,

$$\dot{\varepsilon}_m = a\sigma^{n-2} \exp\left(-\frac{Q_p}{RT}\right) \qquad (12.11)$$

where Q_{SD} is the activation energy for self-diffusion and Q_p is activation energy for pipe-diffusion. The stress exponent, n, also depends on the deformation mechanism. Relationships between the stress exponent and creep deformation mechanism are listed in Table 1. Thus, an understanding of the precise creep mechanism is very important in order to establish an appropriate constitutive equation.

The previous paragraphs have considered situations dominated by a single creep mechanism. If creep involves different mechanisms, the relationship between strain rate and stress does not fit a simple power law as shown in Fig. 12.3 (3). In this figure, the Zener-Hollomon parameter, Z, given by Eq. 12.13 and a normalized stress are used (4).

Table 1 Stress exponent, grain size parameter and activation energy relevant to deformation mechanisms

	n	p	D or Q
Low tem. Power Law	$n+2$	0	pipe
High temp. Power Law			
(Pure metal)	5	0	lattice
(Solid Solution)	3	0	lattice
(precipiation or dispersi			
strengthening)	>7	0	lattice
Grain boundary sliding	2	2	grain boundary
Diffusion Creep			
(grain boundary)	1	3	grain boundary
(lattice)	1	2	lattice

$$Z = \dot{\varepsilon}_m \exp\left(-\frac{Q}{RT}\right)$$

(12.12)

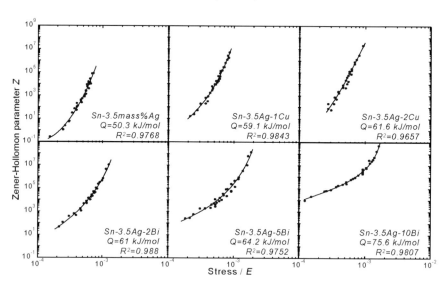

Figure 12-3 Zener-Hollomon parameter, Z, as a function of normalized stress for lead-free solders. R^2 is the squared multiple correlation co-efficient (3)

The power law equation can be rewritten with the Zener-Hollomon parameter as,

$$Z = A\left(\frac{\sigma}{E}\right)^n \qquad (12.13)$$

This equation does not fit the data shown in Fig. 12.3. However, the data may be described by an equation using sinh-function.

$$Z = A(\sinh((a\sigma)/E))^n \qquad (12.14)$$

where the exponent, n, does not correspond to the exponent in Eq 12.10 and 12.11. Figure 12-4 shows Zener-Hollomon parameter, Z, plotted against $\sinh((\alpha\sigma)/E)$ for lead-free solders (4). As can be seen, Equation 12.14 describes well the strain rate-stress behaviour. Equation 12.15 is the more general expression for the constitutive equation for creep using a sinh-function.

$$\dot{\varepsilon}_m = A(\sinh(\alpha\sigma))^n \exp\left(-\frac{Q}{RT}\right) \qquad (12.15)$$

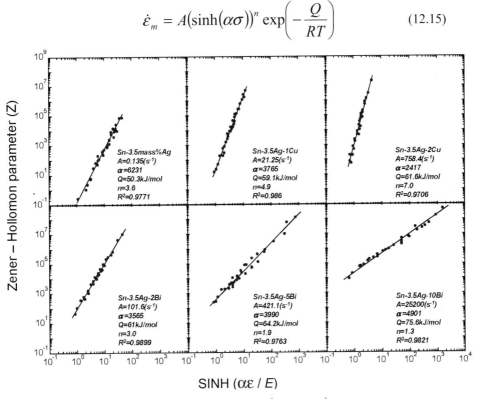

Figure 12-4 Zener-Hollomon parameter, Z, v $\sinh((\alpha\sigma)/E)$ for lead-free solders (3)

It should be recalled that the constitutive equations discussed above apply only for the secondary, or minimum, creep rates. In the strictest sense, they are not complete constitutive equations although they are useful for consolidation of secondary creep data. If primary creep has to be considered to describe creep behaviour, the constitutive equation becomes more complex.

Darvaux et al (5) have produced a fully-developed expression for the creep strain in several solder alloys in joint configurations. The initial inelastic shear strain, xx , was given by;

$$\gamma_o = C \left(\frac{\tau}{G} \right)^m \tag{12.16}$$

where C and m are constants, and G and the temperature-dependent modulus. The primary strain, xx , was obtained from

$$\gamma_p = \dot{\gamma}_s t + \gamma_T \left[1 - \exp(-B\dot{\gamma}_s t) \right] \tag{12.17}$$

where $\dot{\gamma}_s$ is the steady state creep rate, γ_T is the transient creep strain and B is the transient creep coefficient. The steady state creep strain, γ_s, is derived from the product of the time, t, and one of the rate expressions cited previously, e.g.

$$\dot{\gamma}_s = C \left[\sinh(2\tau) \right]^n \exp \left(\frac{-Qa}{RT} \right) \tag{12.18}$$

where Q_a is the apparent activation energy.

In solder alloys with a very small yield strength, the initial strain is usually a combination of elastic and inelastic deformation, and its magnitude depends upon loading rate.

3.3 Cyclic behaviour

3.3.1 Plastic strain based

The Coffin-Manson empirical law, based upon plastic strain range, is the most widely used approach (6, 7).

$$\Delta\varepsilon_p N_f^{\ a} = C \tag{12.19}$$

where N_f is number of cycles to failure, $\Delta\varepsilon_p$ is plastic strain range, α is the fatigue coefficient and C is fatigue ductility. The number of cycles to failure has been defined in many ways (see Chapter 4). The fatigue ductility, C, is approximately equal to the true fracture ductility in tension, ε_f, and the fatigue ductility exponent, C, is around 0.5 for most metals. Generally, a strain-controlled fatigue test is performed on bulk samples under tension-compression cycling to determine the constants. In practice, fatigue cracks initiate in the stress or strain concentration regions of the joints. Due to the complex joint geometry, FEA modelling is best used to determine the maximum equivalent plastic strain equal to the strain in tension. The value for this is then utilised in the Coffin-Manson expression.

This approach to life prediction may be modified to accommodate time-dependent creep phenomena by introducing a frequency term in Equation 12.20 (8).

$$N_f = \left(\frac{C}{\Delta\varepsilon_p} \right)^{\frac{1}{\alpha}} v^{1-k} \tag{12.20}$$

$$\frac{1}{v} = t_{cy} + t_h \tag{12.21}$$

where C, k and α are material constants. The frequency, v, is determined by Equation 12.21. t_{cy} is the time for a cycle and t_h is dwell in the cycle time. Equation 12.20 is known as the frequency-modified Coffin-Manson law. This model is limited in its use for creep-fatigue life prediction, since it is unable to account quantitatively for the effects of loading profile and creep. During thermal cycling, the strains developed at peak temperature and those developed at the minimum temperature are not the same. For example, if the minimum temperature is low enough, little or no creep may be developed during cooling. This suggests that the direction of creep (i.e. tensile creep or compression creep) will influence fatigue damage. To account for the symmetry of the loading profile, Equation 12.20 has been modified using the equivalent plastic strain, $\Delta\varepsilon'_p$, (9).

$$N_f = \left(\frac{C}{\Delta\varepsilon'_p} \right)^{\frac{1}{\alpha}} \left(\frac{v_t}{2} \right)^{1-k} \tag{12.22}$$

$\Delta\varepsilon'_p$ is defined by

$$\Delta\varepsilon'_p = \Delta\varepsilon_p A\left[\frac{\left(v_c/v_t\right)^{kI'}+1}{2}\right]^{\frac{a}{a'}}$$

(12.23)

where $v_t = 1/t_t$ (t_t is time for the deformation in tension slide) and $v_c = 1/t_c$ (t_c is time for the deformation in the compression side of the cycle).

Applicability of strain-based life prediction models

The general form of the Coffin-Manson empirical equation for low cycle fatigue, equation 12.19, successfully describes the behaviour of many solder alloys during *continuous and symmetrical* strain cycling when the strain rates in each direction are similar and there are no dwell periods in the cycle. However, the constants may vary significantly.

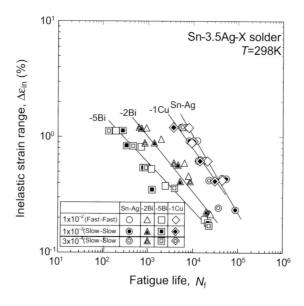

Figure 12-5 Coffin-Manson plots for Sn-3.5Ag-X solder for several strain rates (10)

Figure 12-5 illustrates this for Sn-3.5Ag based solder alloys (10). The data were obtained under total strain controlled fatigue at room temperature at several strain rates. Strain rate does not dramatically affect the fatigue lives of any of the alloys tested, and the endurances after slow-slow cycling are almost the same as those in fast-fast cycling, suggesting that creep damage (or time-dependent damage) is not significant under these conditions. However, departure from symmetry in the strain cycle can result in significant and variable effects on fatigue life and the applicability of the Coffin-

Manson equation. For example, the fatigue lives of Sn-3.5Ag eutectic and Sn-3.5Ag-1Cu alloys are very sensitive to the strain-time profile of the cycle. The endurance after asymmetrical cycling is approximately one-tenth of that in symmetrical strain cycling (10). (Figure 12-6) This dramatic reduction indicates that either tensile creep or compression creep which is not fully reversed has a deleterious effect upon fatigue life, whereas creep does not contribute to the degradation of fatigue endurance when it is imposed in both tension and compression.

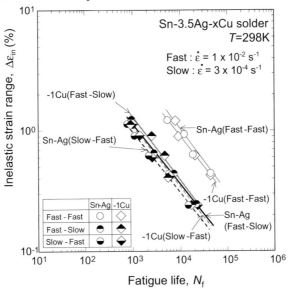

Figure 12-6 Coffin-Manson plots for Sn-3.5Ag-xCu solder during several types of strain cycling (10)

Lead-free Sn-3.5Ag-Bi alloys also exhibit a sensitivity to the symmetry of the strain cycle although the deleterious effect of the creep component upon fatigue damage is different to that in Sn-3.5Ag eutectic and Sn-3.5Ag-1Cu alloys (10). Fatigue lives after slow-fast cycling, that experience a creep component in tension, are approximately one-fifth of those in fast-fast cycling that involves essentially time-independent inelastic deformation (Fig. 12.7). However, in fast-slow cycling, which imposes a creep component only in compression, the fatigue lives are very similar to those observed after fast-fast cycling. Hence, it appears that tensile creep has a deleterious effect on the fatigue lives of bismuth-containing alloys, whereas compression creep is not damaging. This finding is similar to results on creep-fatigue interactions in steel or other structural materials (11–13), and has been termed 'tensile-dwell sensitivity' (14). Other structural alloys, such as nickel and titanium-base gas turbine alloys are generally not sensitive to unbalanced tensile dwells, and may be vulnerable to compression-only hold periods.

In summary, the fatigue behaviour of a relatively small sub group of materials, such as lead-free solder alloys, is not uniform, and strongly depends on the strain-time profile. As the Coffin-Manson empirical equation does not consider the strain profile effects, it is limited for these applications.

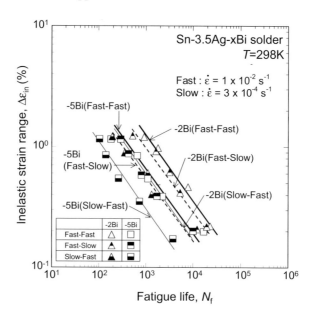

Figure 12-7 Coffin-Manson plots for Sn-3.5Ag-xBi solder during several types of strain cycling (10)

3.3.2 Strain range partitioning

For creep-fatigue life predictions, the strain range partitioning method can quantitatively evaluate inelastic strain ranges by considering the symmetry of cyclic deformation. (15,16). The basic assumption is that the entire reversed inelastic strain range can be partitioned into four generic components identified with creep and plasticity. The components of strain in the tensile half of the cycle are reversed by the compressive half of the cycle: in Figure 12-8a time-independent plasticity is reversed by time-independent plasticity (PP, $\Delta\varepsilon_{pp}$); Figure 12-8b shows time-independent plasticity reversed by creep (PC, $\Delta\varepsilon_{pc}$); Figure 12-8c illustrates creep reversed by time-independent plasticity (CP, $\Delta\varepsilon_{cp}$) and in Figure 12-8d creep is reversed by creep (CC, $\Delta\varepsilon_{cc}$).

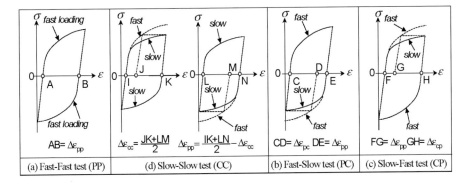

Figure 12-8 Typical hysteresis curves for different strain waveforms (10)

As variations in the component type result in differences in life, even if the magnitude of the strain range is same, four generic strain range-life relationships are determined in a series of tests that feature those four types of components. The relationships may be written in the form,

$$\Delta \varepsilon_{ij} = A_{ij} N_{ij}^{-m_{ij}} \quad (i,j = p, c) \tag{12.24}$$

where N_{ij} is the cyclic life for the partitioned strain range, $\Delta \varepsilon_{ij}$, of interest, and m is the slope of Coffin-Mason plot. Figure 12-9 shows the four generic strain range-life relationships for a Sn-3.5Ag-2Bi alloy (10).

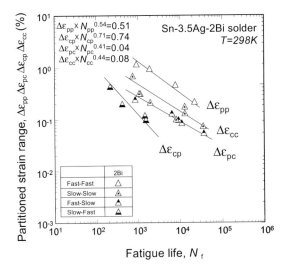

Figure 12-9   Relationship between partitioned strain range and N_f of Sn-3.5Ag-2Bi solder (10)

If the generic strain range-life relationships are known, the fatigue lives can be predicted for any type of inelastic strain cycling by following Miner's linear superposition damage rule.

$$\frac{1}{N_f} = \frac{1}{N_{pp}} + \frac{1}{N_{cc}} + \frac{1}{N_{cp}} + \frac{1}{N_{pc}} \tag{12.25}$$

The SRP method provides precise life prediction for lead-free solder alloys, although it requires many fatigue tests to obtain the four generic strain range-life relationships.

Application of strain range partitioning approach

Strain range partitioning can provide good life prediction when a strain profile sensitivity exists. The partitioned strain range components $\Delta\varepsilon_{pp}$, $\Delta\varepsilon_{cc}$, $\Delta\varepsilon_{cp}$ and $\Delta\varepsilon_{pc}$ versus life relationships for Sn-3.5Ag alloys are shown in Fig. 12.10 (10) that uses the same data as those shown in Figs. 12.5 to 12.7. $\Delta\varepsilon_{pp}$ is defined as the inelastic strain range in a fast-fast test, when subjected to a strain rate of $10^{-2}s^{-1}$, which is sufficiently high to exclude creep. $\Delta\varepsilon_{cp}$, $\Delta\varepsilon_{pc}$ and $\Delta\varepsilon_{cc}$ are determined from the hysteresis loops obtained in a series of tests (slow-fast, fast-slow and slow-slow) that feature the three generic inelastic components, as discussed above.

The data on each partitioned inelastic strain range versus fatigue life fit the Coffin-Manson equation. Hence, using the damage rule, the fatigue life can be expressed as follows:

For Sn-3.5Ag alloy:

$$\frac{1}{N_f} = \left(\frac{\Delta\varepsilon_{PP}}{1.06}\right)^{1.92} + \left(\frac{\Delta\varepsilon_{CP}}{1.92}\right)^{1.67} + \left(\frac{\Delta\varepsilon_{PC}}{4.76}\right)^{0.98} + \left(\frac{\Delta\varepsilon_{CC}}{0.24}\right)^{1.56} \tag{12.26}$$

For Sn-3.5Ag-2Bi alloy:

$$\frac{1}{N_f} = \left(\frac{\Delta\varepsilon_{PP}}{1.51}\right)^{1.85} + \left(\frac{\Delta\varepsilon_{CP}}{0.74}\right)^{1.41} + \left(\frac{\Delta\varepsilon_{PC}}{0.04}\right)^{2.43} + \left(\frac{\Delta\varepsilon_{CC}}{0.08}\right)^{2.27} \tag{12.27}$$

For Sn-3.5Ag-1Cu alloy:

$$\frac{1}{N_f} = \left(\frac{\Delta\varepsilon_{PP}}{1.47}\right)^{1.82} + \left(\frac{\Delta\varepsilon_{CP}}{140.2}\right)^{0.71} + \left(\frac{\Delta\varepsilon_{PC}}{565.6}\right)^{0.65} + \left(\frac{\Delta\varepsilon_{CC}}{7.46}\right)^{1.22} \tag{12.28}$$

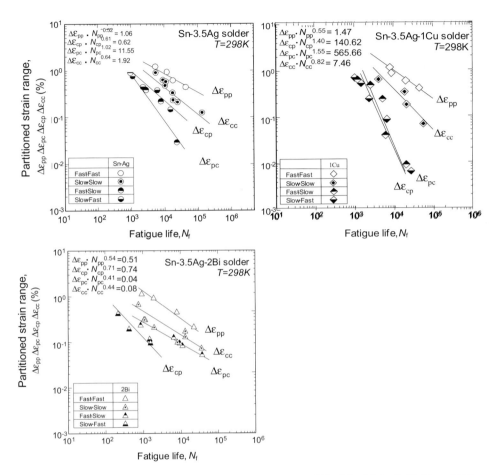

Figure 12-10 Relationship between partitioned strain range and N_f of Sn-3.5Ag-X (X=Bi and Cu) solder (10)

Agreement between measured and predicted lives is good, generally within a factor of two. (Figure 12-11).

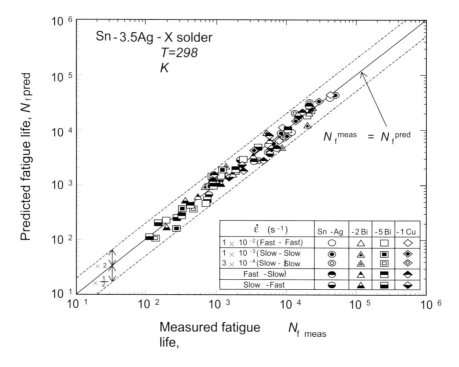

Figure 12-11 Comparison between measured and predicted fatigue life using the strain range partitioning approach (10)

3.3.3 Energy-based models

Fatigue life can be correlated with the mechanical energy of the hysteresis loop, expressed in the form,

$$N_f^{\ m}W = C_E \qquad (12.29)$$

where W is the mechanical hysteresis energy density, m and C_E are constants. This approach recognizes that stress is required to move the dislocations necessary for irreversible local plastic deformation, and has often been employed for solder alloys (17). Bisego et al. (18) found the following relationship between absorption energy during one cycle and number of cycles to failure.

$$eN_f = Q_1 \qquad (12.30)$$

where e is the absorption energy per mol during one cycle and Q_l is the work load to failure. e is determined from the area of stress-strain hysteresis loop. Q_l is a function of strain rate and is determined from (19)

$$Q_1 = A\varepsilon^q \tag{12.31}$$

where A is material constant. If Q_l is known, the number of cycles to failure can be obtained by measuring the area of hysteresis loop for any condition. This approach has not been applied to solder alloys yet, although it appears to offer significant potential.

Ostergren (20) proposed $\sigma_{max}\Delta\varepsilon_{in}$ as the damage function, and found the following relationship between the fatigue life and the function.

$$\sigma_{max}\Delta\varepsilon_{in}N_f^{\Phi} = C \tag{12.32}$$

where σ_{max} is the maximum value of stress, $\Delta\varepsilon_{in}$ is the inelastic strain range and ϕ and C are constants; σ_{max} is given in following form,

$$\sigma_{max} = \sigma_m + \Delta\sigma/2 \tag{12.33}$$

where σ_m is the mean stress. The basic assumption of this model is that the fatigue damage can be correlated with the energy in tension side of the hysteresis loop. When the mean stress is zero, it becomes a similar to the Coffin-Manson expression.

If the time-dependent damage is considered, a frequency term can be added to Equation 12.32 (20).

$$\sigma_{max}\Delta\varepsilon_{in}N_f^{\Phi}v^{\Phi(k-1)} = C \tag{12.34}$$

where k and C are constants.

If k=1, there are two types of time-dependent damage: (a) If the damage is independent of the symmetry of loading, v is defined as follows,

$$v = \frac{1}{t_{cy}} = \frac{1}{t_0 + t_t + t_c} \tag{12.35}$$

The cyclic time, t_{cy}, is the sum of ramp time, t_0, tensile hold time, t_t, and compression hold time, t_c.

(b) If the damage depends on the symmetry of loading, v is defined as follows,

$$v = \frac{1}{t_0 + t_t + t_c}, \quad t_t \geq t_c \tag{12.36}$$

$$v = \frac{1}{t_0} t_i \langle t_c \tag{12.37}$$

The model is simple and very effective in predicting fatigue life under creep-fatigue conditions for steels, although its applicability to solder alloys requires further confirmation.

Application of energy-based life prediction models

The hysteresis energy density is defined as the area of the hysteresis loop divided by molecular number of gauge section. (21). Again, analysing the data shown in Figures 12.5 to 12.7, a linear relationship between the logarithm of fatigue life and the logarithm of hysteresis energy density, indicates that fatigue life can be predicted from equation 12.29 (22): where W is the mechanical hysteresis energy density, m and C_E are constants. Figures 12.12 and 12.3 demonstrate this linearity for each strain profile. Thus fatigue life depends on the strain profile at any given hysteresis energy density, and indicates that the model cannot be used to predict fatigue life for varying strain profiles.

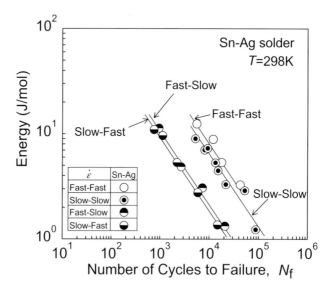

Figure 12-12 Fatigue life as a function of hysteresis energy for Sn-3.5Ag (22)

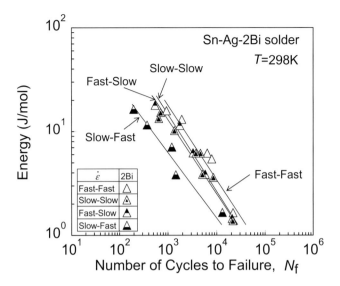

Figure 12-13 Fatigue life as a function of hysteresis energy for Sn-3.5Ag-2Bi solder alloy (22)

To account for the effect of frequency, a frequency modified form of the expression has been proposed (23, 24) as follows:

$$\left[N_f v^{(k-1)} \right]^m \frac{W}{v^n} = C_{E1} \tag{12.38}$$

where k is a frequency exponent, determined from the relationship between fatigue life and frequency, and m is another constant, determined from the relationship between strain energy density and frequency.

The model cannot accommodate the effect of different strain profiles. Further, the constant C_{E1} depends on temperature, and the model may be in error where creep effects are dominant. More modifications will be required to predict creep-fatigue life of lead-free solder alloys.

The *strain energy partitioning model (SEP)* synthesizes the characteristics of the SRP and Ostergren's damage function as follows (25):

$$N_{ij} = C_{ij} \left(\sigma_{max} \cdot \Delta \varepsilon_{ij} \right)^{D_{ij}} (i, j + p, c) \tag{12.39}$$

where N_{ij} is the cyclic life for the partitioned strain range $\Delta \varepsilon_{ij}$ of interest, σ_{max} is maximum value of stress and D_{ij} is a constant. Figures 12.14 and 12.15 show that SEP describes the effect of different strain profiles as well as the SRP.

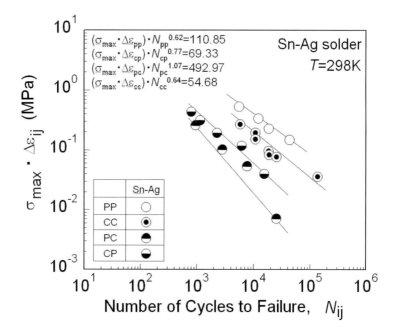

Figure 12-14. Fatigue life as a function of partitioned strain energy for a Sn-Ag alloy (22)

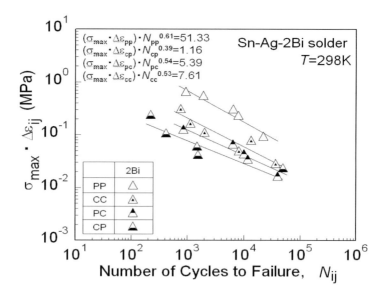

Figure 12-15. Fatigue life as a function of partitioned strain energy for a Sn-Ag-Bi alloy (22)

4   CONCLUSIONS

Finite element analysis is necessary to obtain the stress-strain relationships that exist in joints due to their complex geometry. Constitutive equations for the operative deformation mechanism are required for input, and often the expression employed refers to a steady state phase which may describe only a small portion of the total process. Many fatigue life prediction models are available. However, most were developed to predict the fatigue life for high temperature structural alloys, and the extent to which these approaches can be applied to solder alloys has yet to the fully determined. Many models cannot accurately describe the effect of different strain profiles. Lead-free alloys exhibit various responses to strain cycling with dwells. Some are sensitive to unbalanced dwell-containing cycles, whereas others experience a life deficit only after tensile-dwell cycling. Strain range partitioning methods can predict behaviour under creep-fatigue conditions for some solder alloys although extensive experimental testing is involved in this method. The hysteresis energy density approaches for life prediction are unable to accommodate dwell-containing or asymmetric cycles.

5   REFERENCES

1   TG Nieh, J Wadsworth and OD Sherby, Superplasticity in Metals and Ceramics, Cambridge University Press (1997), p.32.
2   JG Harper, LA Shepard and JE Dorn, Acta Metal., 1958, 6,.509.
3   Y Kariya, T Morihata, E Hazawa and M Otsuka., Proc. of 6th symposium on Microjoining and Assembly Technology in Electronics, Japanese Welding Soc., 2000, p.281.
4   C Zener and JH Hollomon, J. Appl. Phys., 1944, 15, 22.
5   R Darvaux, K Banerji, A Mawer and G Dodd, Ball Grid Array Technologies, (Ed JH Lau), McGraw Hill, 1995, Chap 13, 379.
6   SS. Manson, NASA, National Aeronautics and Space Administration, TN2933 (1953).
7   LF Coffin, Jr., Trans. ASME, 1954, 76, 931.
8   LF. Coffin, Proc. of Second International Fatigue Conference on Fracture, 1969, p.643.
9   LF. Coffin, Proc. of Inst. Mech. Eng., 1974, 188,109.
10   Y Kariya, T Morihata, E Hazawa and M Otsuka, J. of Electronic. Materials, 2001, 30, 1184.
11   R Viswanathan, Damage Mechanics and Life Assessment of High Temperature Components, ASM International, 1989, p.171.
12   R Viswanathan, JR Foulds and DA Roberts, Proc. International Conference on Life Extention and Assessment, Hague, 1988.
13   BJ Cane, PF Aplin and JM Brear, J. Pressure Vessel Tech., 1985, 107, 295.
14   WJ Plumbridge, High Temperature Fatigue: Properties and Prediction (Ed R. P. Skelton) Elsevier, (London and New York) 1987, Chap 4.
15   SS Manson,GR Halford and MH Hirschberg Design for Elevated Temperature Environment (ed. S. Y Zamrik) ASME, 1971, p.12.
16   GR Halford and SS Manson, ASTM STP 612, 1976, 239.
17   For example, HD Solomon and ED Tolksdorf, Trans. ASME., J. of Electronic Packaging, 1995, 117, 130.
18   V Bisego et al., An Energy Based Criterion for Low Cycle Fatigue Damage Evolution, in Material Behaviour at Elevated Temperatures and Component Analysis, Book. No. H00217, PVP 60, ASME (1982).
19   C Bartoloni and G Ragazzoni, Proc. of Second International Fatigue Conference on Creep and Fracture of Engineering Materials and Structures, p.1029.
20   WJ Ostergren, ASME J. Test. Eval., 1976, 4, 327.
21   JD Morrow, ASTM STP 378, 1964, p.45.

22 Y. Kariya, T Morihata and M Otsuka, Proc. Symp. On Microjoining and Assembly Technol. In Electronics, 2002, 8, 437.
23 HD. Solomon and E. D. Tolksdorf, ASME J. Electron. Packaging, 1996, 118, 67.
24 XQ. Shi, HLJ Pang, W Zou and ZP Wang, Scripta Materia, 1999, 41, 289.

CHAPTER 13

LIFE PREDICTION BY CRACK PROPAGATION APPROACH

1 INTRODUCTION

As discussed previously, the Coffin-Manson approach to fatigue life prediction involves an assessment of strength reduction during strain controlled cycling. An empirical law is developed between fatigue life (as defined by a fall-off in strength) and the applied strain range. No mechanistic relationship is directly involved because several processes (cyclic hardening/softening, microstructural coarsening/refinement and multiple crack initiation and growth) may all contribute to the measured load change. An alternative approach to determine the fatigue life is by the *crack propagation model*. This addresses the question raised in Chapter 3 (S4.2) regarding how quickly (or in how many cycles) does a crack grow from its initial size to a dimension that causes failure. The initial defect size is either that which is observed by non-destructive examination or the resolution of the technique being employed to seek cracks. The final crack length may be determined from toughness measurements, or be regarded as the dimension of the solder joint. If the crack growth rate is known as a function of the crack length, applied stress or strain range, and other relevant parameters, then the fatigue life can be determined by integrating the crack growth expression between the initial and final crack lengths.

The fatigue life of a cyclically loaded component is composed of both crack initiation and crack propagation stages. At high strain ranges (low cycle fatigue) crack initiation is rapid and propagation is dominant in determining endurance. The fatigue life after low strain range cycling is mainly spent in initiating a crack, so life estimation in this case has a significant element of conservatism built in. Any feature that acts as a stress concentrator promotes early initiation. In most metals and alloys under cyclic loading, catastrophic failure is preceded by a substantial amount of stable crack propagation. In this chapter, the basics of the crack propagation approach to determine fatigue life, and their application to solders will be discussed.

2 BASICS OF THE CRACK PROPAGATION APPROACH

Fatigue life is governed by the conditions prevailing at the crack tip which determine its growth. The various situations are now considered.

2.1 Linear elastic fracture mechanics (LEFM) conditions

The rate of growth of a fatigue crack length, a, subjected to a constant stress range, $\Delta\sigma$, is expressed in terms of the crack length increment per cycle, da/dN. Values of da/dN for different loading conditions are determined from experimentally measured changes in crack length over a number of elapsed fatigue cycles. When the applied stress range is held constant, the rate of growth of a fatigue crack during Mode 1 (fluctuating tension or tension-compression) loading generally increases with increasing number of fatigue cycles and crack length. Figure 13-1 schematically illustrates typical fatigue crack growth curves.

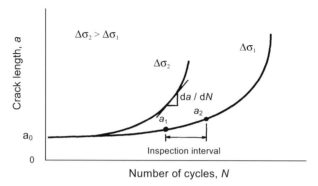

Figure 13-1. Typical crack growth behaviour in constant amplitude fatigue loading

When the cyclic stresses applied to a component are small, the zone of plastic deformation ahead of the fatigue crack is surrounded by an elastic field. Linear elastic fracture mechanics solutions provide appropriate descriptions for fatigue fracture. Characterisation of the rate of fatigue crack growth is based on the *stress intensity factor range*, ΔK,

$$\Delta K = K_{max} - K_{min} \qquad (13.1)$$

K_{max} and K_{min} are the maximum and minimum values, respectively, of the stress intensity factor during a fatigue stress cycle. For an edge-cracked fatigue test specimen, in fluctuating tension

$$K_{max} = Y\sigma_{max}\sqrt{\pi a}, \quad K_{min} = Y\sigma_{min}\sqrt{\pi a}, \qquad (13.2)$$

where Y is a geometrical factor which depends on the ratio of crack length, a, to the width of the specimen W, and σ_{max} and σ_{min} are the maximum and minimum values, respectively, of the fatigue stress cycle. For reversed cycling, when compressive loads are applied, ΔK, is taken either as the tensile component, assuming the minimum stress is zero, or as the difference between the maximum stress intensity level and that associated with crack opening.

The fatigue crack growth increment, da/dN, is related to the stress intensity factor range by the power law relationship (1),

$$\frac{da}{dN} = C(\Delta K)^m \tag{13.3}$$

where C and m are scaling constants. These constants are influenced by such variables as material microstructure, environment, temperature and load ratio, R, defined as

$$R = \frac{\sigma_{min}}{\sigma_{max}} = \frac{K_{min}}{K_{max}} \tag{13.4}$$

The exponent, m, is typically between two and four for ductile metals. Equation 13.3 is applicable for a single mode of far-field loading and for a fixed value of R. For tensile fatigue, ΔK refers to the range of Mode I stress intensity during the stress cycle. Similarly, a stress intensity factor range, ΔK_{II} or ΔK_{III}, can also be used to characterize fatigue crack growth in Mode II or Mode III, respectively. Experimental data gathered for a wide range of metallic materials confirm a power law relationship of the form given in Equation 13.3. If environmental conditions, including temperature, have a strong effect on fatigue fracture, loading parameters such as cyclic frequency and waveform would also be expected to have a pronounced effect on crack growth rates. Changes in these test conditions can lead to changes in the empirical constants C and m and in the fatigue crack growth rates.

With this approach, the fatigue life or the number of fatigue cycles to failure, may be calculated by integrating Equation 13.3 from an assumed initial crack size, a_o, to a critical crack size, a_f.

$$\frac{da}{dN} = C\left(Y\Delta\sigma\sqrt{\pi a}\right)^m \tag{13.5}$$

If fatigue loading involves constant amplitude of far-field stresses and a crack length change $(a_f - a_o)$ over which Y is roughly constant, then

$$CY^m(\Delta\sigma)^m \pi^{m/2} \int_0^{N_f} dN = \int_{a_o}^{a_f} \frac{da}{a^{m/2}} \tag{13.6}$$

The fatigue life, N_f, is given by

$$N_f = \frac{2}{(c-2)CY^m(\Delta\sigma)^m \pi^{m/2}} \left\{ \frac{1}{(a_0)^{(m-2)/2}} - \frac{1}{(a_f)^{(m-2)/2}} \right\} \tag{13.7}$$

For $m=2$,

$$N_f = \frac{2}{CY^2(\Delta\sigma)^2 \pi} \ln = \frac{a_f}{a_0} \qquad (13.8)$$

When the value of m does not equal 2, numerical integration is required to determine the fatigue life.

2.2 Elastic-plastic fracture mechanics (EPFM) conditions (The J-integral)

The stress intensity factor, K, provides a unique characterisation of the near-tip fields under small-scale yielding conditions. The corresponding loading parameter for the characterisation of monotonic, non-linear fracture in rate-independent materials is the *J*-integral (2). Although some of the features of the J-integral were embedded in energy concepts, the particular form of this line integral has led to the unifying theoretical basis for nonlinear fracture mechanics.

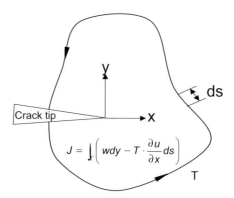

Figure 13-2. A contour around a crack tip and the nomenclature used in the definition of the J-integral

Consider a cracked body subjected to a monotonic load. Assuming that the traction T is independent of crack size and that the crack faces are traction-free, the line integral, J, along any contour T that encircles the crack tip is given by

$$J = \int \left(wdy - T \cdot \frac{\delta u}{\delta x} \right) ds \qquad (13.9)$$

where u is the displacement vector, y is the distance along the direction normal to the plane of the crack, s is the arc length along the contour, T is the traction vector and w is strain energy density of material. The stresses are related to w by the relationship

$\sigma = \partial w / \partial \varepsilon$. For a material which is characterised by linear or non-linear elastic behaviour, J is independent of the path taken to compute the integral.

Although conditions of non-proportional loading and the occurrence of elastic unloading appear to violate the fundamental basis on which the application of J-integral to fracture problem is predicated, Dowling and Begley (3) have proposed a power law characterization of fatigue crack advance under elastic-plastic conditions based on the cyclic J-integral, ΔJ, during a fatigue cycle as follows,

$$\frac{da}{dN} = C' \left(\Delta J_t \right)^{m'} \qquad (13.10)$$

where C' and m' are scaling constants. ΔJ_t consists of elastic part, ΔJ_e, and plastic part, ΔJ_p, and the total ΔJ_t is given by

$$\Delta J_t = \frac{\Delta K^2}{E} + \Delta J_p \qquad (13.11)$$

where E is the elastic modulus. The stress and strain relationship for the elastic-plastic material can be expressed in the Ramberg-Osgood model as

$$\varepsilon = \frac{\sigma}{E} + \left(\frac{\sigma}{a} \right)^{\frac{1}{\beta}} \qquad (13.12)$$

where α is a material constant and β is the strain hardening coefficient ($0<\beta<1$). Equation 13.11 can be expressed for surface cracked specimens according to Dowling as follows (4),

$$\Delta J_t = \left(\frac{3.2 \Delta \sigma^2}{2E} + \frac{5 \Delta \sigma \cdot \Delta \varepsilon_p}{1 + \beta} \right) \cdot a \qquad (13.13)$$

This approach provides good characterisation of the growth of short fatigue cracks of length comparable to the near-tip plastic zone size, and longer fatigue flaws in almost fully yielded specimens, in some materials under certain cyclic loading conditions.

2.3 Crack propagation under fully plastic conditions

If a specimen is subjected to a large deformation, such as thermal cyclic loading, damage by elastic deformation may be negligible. So, Equation 13.13 becomes

$$\Delta J_t = \Delta J_p = \left(\frac{5\Delta\sigma \cdot \Delta\varepsilon_p}{1+\beta} \right) \cdot a \tag{13.14}$$

The relation between the stress increment and the strain increment during plastic deformation is given by

$$\Delta\sigma = a \cdot \left(\Delta\varepsilon_p \right)^{\beta} \tag{13.15}$$

If $A = \dfrac{5\alpha}{1+\beta}$ and $n = \beta + 1$, then Equation 13.14 can be rewritten as

$$\Delta J_t = A \cdot \left(\Delta\varepsilon_p \right)^{n} \cdot a \tag{13.16}$$

where A and n are materials constants ($1 < n < 2$). By substituting Equation 13.16 into Equation 13.10, the crack growth rate becomes

$$\frac{da}{dN} = C'' \cdot \left(\Delta\varepsilon_p^n \cdot a \right)^{m'} \tag{13.17}$$

where C'' is a material constant and m' for ductile materials is 1. The number of fatigue cycles to failure is calculated by integrating Equation 13.17 from an assumed initial crack size a_o to a critical crack size a_f.

$$N_f = \frac{1}{YC''\Delta\varepsilon_p^n} \ln \frac{a_f}{a_0} \tag{13.18}$$

where Y is a geometrical factor which depends on the ratio of crack length, a, to the width of the specimen.

2.4 Crack propagation under creep-fatigue conditions

At temperatures that are significant fractions of the homologous temperature, the mechanisms responsible for fracture become time-dependent as well as cycle-dependent. Time-dependent processes may be controlled by creep and/or environment. For time-dependent fatigue crack propagation, the stress intensity factor range is still the controlling parameter if the zone of inelastic deformation at the crack tip is small in size compared with the crack length and the size of the uncracked ligament. i.e. LEFM conditions apply. However, at high temperatures and low test frequencies, the mechanical fatigue and time-dependent creep components of crack growth may be linearly superposed to derive an overall crack extension rate. The total fatigue crack growth rate is then given by (5)

$$\left(\frac{da}{dN}\right) = \left(\frac{da}{dN}\right)_F + \left(\frac{da}{dN}\right)_{CR} \tag{13.19}$$

where the subscripts F and CR denote the contributions from fatigue and creep, respectively. Similar partitioning methods and their various empirical adaptations have also been employed in the context of thermomechanical fatigue. The fatigue component of high temperature crack growth is then characterized by Equation 13.3, 13.10, or 13.17. Similarly, the time-dependent component $(da/dN)_{CR}$ is characterized by (5)

$$\frac{da}{dt} = C_3 \left(P_{CR}\right)^{m_3} \tag{13.20}$$

where C_3 and m_3 are material constants. P_{CR} is the creep fracture parameter which is chosen to be C^* for large-scale deformation. C^* for ductile materials can be estimated by

$$C^* = \frac{P\dot{V}}{BW} \left[\frac{n}{n+1}\right] \left(\frac{2}{1-a/W} + 0.522\right) \tag{13.21}$$

where P is the applied point load, $\dot{V}$ is the load-line displacement rate, B, W and a are the specimen thickness, width and crack length respectively, and n is the creep exponent for a compact tension specimen (6).

One method for converting da/dt into $(da/dN)_{CR}$ is to use the relationship (6).

$$\left(\frac{da}{dN}\right)_{CR} = \int_0^t C_3 C^{*m_3} dt \tag{13.22}$$

where *t* is time for creep *e.g.* dwell time. This approach, in principle, can also account for the effect of test frequency and waveform on the rates of high temperature crack growth.

In the final section of this Chapter, the extent to which these general analytical approaches can be applied to solders is examined.

3   APPLICATION OF CRACK PROPAGATION METHODS TO SOLDER ALLOYS

Fatigue crack propagation under LEFM conditions in bulk solders is not common because the low yield strength of these alloys favours elastic-plastic or fully plastic situations. However, elastically-dominated behaviour has been reported in bulk material and in joints (7, 8). For both Mode I and Mode II (tensile and shear) loading, expressions of the form, da/dN = B (ΔK_I or ΔK_{II})^m describe the data obtained. The actual crack path is influential in determining the most appropriate analysis to be used. For example, if the crack grows along the solder - IMC interface, more sophisticated modelling would be required. Conversely, if the majority of the fatigue life is spent in extending the crack through the brittle IMC, LEFM would probably suffice. Yet more evidence for the need to understand the operative failure mechanisms!

Since the inelastic strain range is the main driving force for fatigue failure, crack growth models under elastic-plastic, fully-plastic, or fatigue-creep conditions may be used to predict the fatigue life of solder joints. However, the available experimental data are limited. In this section, some of the data for solder alloys will be used to demonstrate fatigue crack characterisation under these conditions.

Zhao et al. (9) have used the approach with the *J*-integral and C^* parameter for data obtained on Pb-5Sn, Sn-37Pb and Sn-3.5Ag solders. Relationships between ΔJ and da/dN for Pb-5Sn solder at various frequencies and stress ratios are shown in Figure. 13.3. Fatigue crack growth rates increase with decreasing frequency while crack growth curves, at various stress ratios, under a frequency of 10 Hz fall into a narrow band. Since a high stress ratio enhances the effect of creep, it is expected that the stress ratio will have a significant influence on the fatigue crack growth behaviour at lower frequencies. The results indicate that the crack growth behaviour transfers from cycle-dependent to time-dependent with decreasing frequency, when the gradient of the graph increases at high values of ΔJ.

The relationship between da/dt and C^* for the creep crack growth test, as well as those for three cases of cyclic crack growth tests, are shown in Figure 13-4 (9). The growth curve for the fatigue crack growth test at 0.01 Hz and *R*=0.1 has the highest coincidence with that for the creep crack growth test. Crack growth rates for the tests at 0.1 Hz under *R*=0.1 and at 10 Hz under *R*=0.7 are about an order of magnitude greater than in the creep crack growth test at similar C^* values.

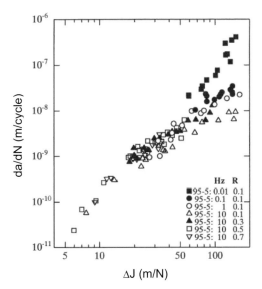

Figure 13-3. Relationship between da/dN and ΔJ at various stress ratios and frequencies for Pb-5Sn solder (CT specimen) (9)

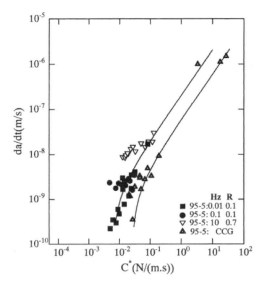

Figure 13-4. Relationship between da/dt and C* for Pb-5Sn solder (9)

The relationship between da/dN and ΔJ for a Sn-3.5Ag solder tested at various frequencies and stress ratios is shown in Figure 13-5 (10). Most data from tests performed under low stress ratios of 0.1 and 0.3 fall into a broad band. However, when R is raised to 0.5 at 10 Hz, the growth rates progressively depart from this band as their values increase due to creep effects. Fatigue crack growth rates become higher with further increasing stress ratio and decreasing frequency.

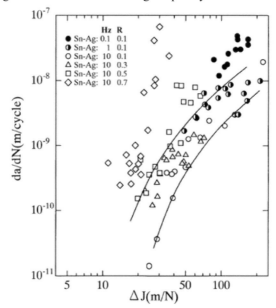

Figure 13-5. Relationship between da/dN and ΔJ in Sn-3.5Ag solder (10)

The relationship between da/dt and C^* for creep crack growth tests, as well as those for cyclic crack growth tests, is shown in Fig. 13.6 (10). At high stress ratios ($R \geq 0.5$), the cyclic crack growth data almost coincide with those for creep crack growth. The data for $R > 0.5$ at 10 Hz fall into the band when the crack growth rates are high but the dominance of time-dependent behaviour is indicated at low growth rates by the broad spread of experimental points.

In summary, these results demonstrate that the fatigue crack growth behaviour of eutectic Sn-3.5Ag solder is cycle dependent at low stress ratios ($R < 0.3$) and high frequencies ($f > 1$ Hz), while it changes to be time dependent with increasing stress ratio and decreasing frequency.

Figure 13-7 shows the da/dN v ΔJ curves for Sn-37Pb and Sn-3.5Ag solder in the cycle-dependent crack growth regime, and Figure 13-8 shows the da/dt and C^* curves in the time-dependent crack growth regime (10). The fatigue crack growth resistance of Sn-3.5Ag solder is higher than that of Sn-37Pb solder in the time-independent regime, although the crack growth resistance of the Sn-3.5Ag solder almost coincides with that for the eutectic Sn-37Pb solder in time-dependent regime. (Figure 13-8).

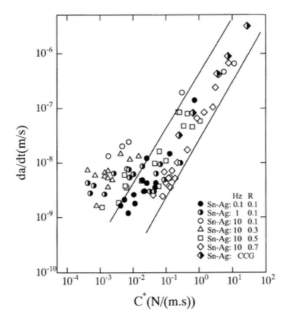

Figure 13-6. The da/dt and C* relationship in Sn-3.5Ag solder (10)

A transition from cycle-dependent to time-dependent crack growth, with increase in stress ratio and reduction in frequency, is common in many solder alloys (11). In the cycle-dependent region, crack growth rates in lead-free alloys are lower than those in Sn-95Pb or Sn-37Pb, and can be expressed as.

$$\frac{da}{dN} = 2.6 \times 10^{-13} \Delta J^{2.1} \tag{13.23}$$

In the lead-containing alloys, the equivalent expression is

$$\frac{da}{dN} = 3.8 \times 10^{-12} \Delta J^{1.9} \tag{13.24}$$

However, the data may be normalized, using an 'effective' stress intensity and Young's modulus, so that a single expression describes fatigue crack growth in all the alloys under all the conditions in the cycle-dependent growth zone (12) i.e.

$$\frac{da}{dN} = 6.83 \times 10^{-12} \left(\Delta K_{eff} / E \right)^{4.05} \tag{13.25}$$

Under time-dependent growth conditions, crack growth data from both lead-free and lead-containing alloys, fall into a narrow band that can be expressed as (11)

$$\frac{da}{dt} = 8.34 \times 10^{-8} C *^{1.08}$$

(13.26)

Ternary Sn-Ag-Cu alloys with bismuth additions are exceptional in that they exhibit only cycle-dependent growth characteristics under the conditions employed. This is attributed to the additional component of strengthening imparted by the bismuth in solid solution (11).

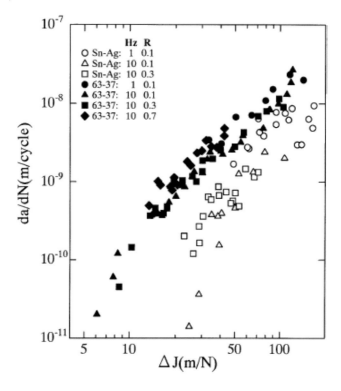

Figure 13-7. The da/dN v ΔJ curves for Sn-3.5Ag and Sn-37Pb solders in the cycle- dependent crack growth regime (10)

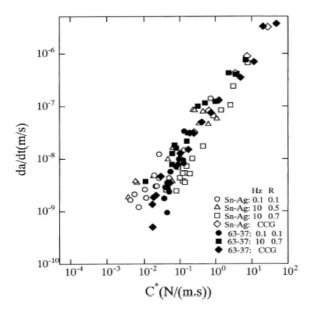

Figure 13-8. Comparison of da/dt v C* curves between Sn-3.5Ag and Sn-37Pb solders (10)

An increase in temperature also promotes time-dependent behaviour. The capabilities of the ΔJ and C^* approaches to accommodate this have been well demonstrated by Kanchanomai (13). Figures 13.9 and 13.10 show the advantage of the C^* analysis in normalizing crack growth data from a range of temperatures.

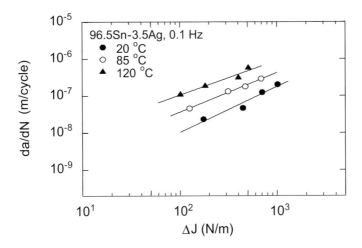

Figure 13-9. Relationship between crack growth rate and ΔJ for different temperatures (13)

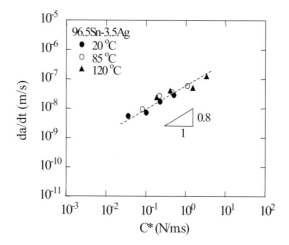

Figure 13-10. Relationship between crack growth rate and C* for different temperatures (13)

Lee *et al* (14) have investigated fatigue crack propagation in Sn-36Pb-2Ag solder as a function of *plastic strain* range. For a given plastic strain range, the logarithmic crack length is proportional to the number of cycles (Figure 13-11) and the crack growth rate is proportional to the logarithmic plastic strain range (Figure 13-12). These results suggest that crack growth for the alloy obeys Equation 13.17 and is governed by the *J*-integral at relatively high frequencies. Thus, crack growth for this solder alloy may be characterised by the *J*-integral (time-independent regime) or *C** parameter (time-dependent regime).

Mode II (shear) fatigue crack growth rates in model joints have been correlated with both ΔJ and ΔW_p, with a power term, b, similar to that observed during Mode1 loading. However, the value of b was considerably higher when the crack growth rates were correlated with stress intensity range (8).

Crack growth measurements made on three-bar model joint specimens during thermomechanical fatigue (30 to 125°C) indicate that the crack growth rate diminishes with continued cycling and increase in crack length (15) (Figure 13-13). This contrasts with the traditional stress-controlled (tensile, Mode1) situation in which the crack propagation rate generally increases with both cycles and crack length. This is a good example of shear fatigue under total strain control, where the driving force for crack growth becomes more remote from the crack tip as it extends into the body of the joint. In these experiments it was possible to correlate the crack growth rate with the cyclic strain energy and the strain range.

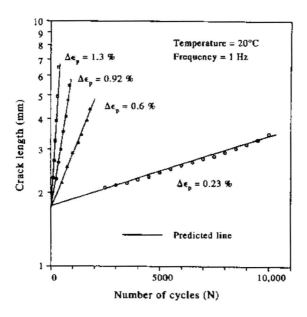

Figure 13-11. Crack length vs. number of cycles for the Sn-36Pb-2Ag at four plastic strain ranges (14)

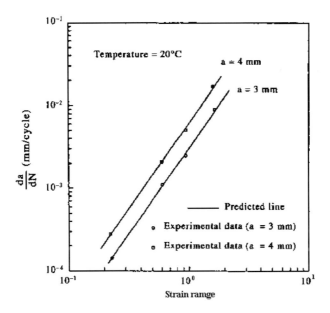

Figure 13-12. Crack growth rate v   plastic strain range for Sn-36Pb-2Ag solder(14)

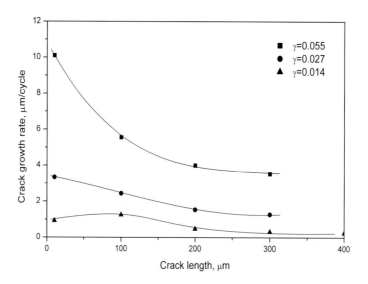

Figure 13-13 Crack growth in a model joint sample as a function of crack length and shear strain range (15)

4 CONCLUSIONS

Fracture mechanics is a potentially valuable tool for life prediction of solder joints, since the underpinning analysis exists for most situations. Crack propagation in solder alloys is generally best characterised by the *J*-integral or a creep fracture parameter, and this approach may be suitable for fatigue life prediction of the solder joints. Experimental data describing crack propagation in solder alloys are limited at present, and more are required to validate the life prediction approach. It is important in joints to identify the location of cracking in order to employ the most appropriate analysis.

5 REFERENCES

1   P Paris and F Erdogen, Trans ASME, 1963, 85, 528.
2   JR Rice, *J. Appl. Mech.* Trans ASME, 1968, 35, 379.
3   NE Dowling and JA Begley, ASTM STP 1967, *590*, 19.
4   NE. Dowling, ASTM STP 637, (1977), 97.
5   A Saxena, RS. Willams, and TT Shih, ASTM STP 743, 1981, 86.
6   A Saxena, Engrg. Fracture Mech. 1991, 40, 721.
7   WA Logsdon, PK Liaw and MA Burke, Engrg. Fracture Mech., 1990, 36, 183.
8   Z Guo and H Conrad, J Electronic Packaging, 1993, 115, 159.
9   J Zhao, Y Miyashita and Y Mutoh, Int. J. Fatigue, 2000, 22, 665.
10 J Zhao, Y Miyashita and Y Mutoh, Int. J. Fatigue, 2001, 23, 723.

11 Y Mutoh, J Zhao, Y Miyashita and C Kanchanomai, Soldering and Surface Mount Technol., 2002, 14, 37.
12 J Zhao, Y Mutoh, Y Miyashita and L Wang, Engrg Fracture Mech., 2003, 70, 2187.
13 C Kanchanomai, 'Study of low-cycle fatigue behavior and mechanisms of lead-free solders', Doctoral/Ref Papers Dissertation, Nagaoka University of Technology, Japan, June, 2002.
14 SB Lee and JK Kim, Int. J. Fatigue, 1997, 9, 85.
15 WJ Plumbridge and XW Liu, to be published.

CHAPTER 14

RELIABILITY AND STATISTICAL ANALYSIS OF DATA

1 INTRODUCTION

In the design of engineering components and structures, two quite different strategies can be adopted with regard to performance. *Will the product fail?* or *When will the product fail?*

There are two types of approach that may be followed to predict life and ensure component reliability.

(1) The *deterministic* approach involves identification of damaging mechanisms and determination of the kinetics of these processes during service. This approach is sometimes called the '*Physics of Failure*' strategy (although 'Physical Metallurgy of Failure' might be more accurate) and has been applied to many engineering applications. Once the kinetics of damage accumulation have been determined, the most deleterious mechanism for a given set of operating conditions may be identified, and estimations of life made.

The required testing may involve closely controlled experimental evaluation of a complete component or elements of it, such as the bulk materials involved, in order to develop constitutive equations which describe material behaviour. In conjunction with stress and strain distributions generated by finite element modelling of the system, and with an appropriate life predictive expression, overall performance may be determined. Relatively few variables are investigated and the approach yields a single, or very narrow spread, of values for failure. Within this text, emphasis has been placed upon thermomechanical fatigue as the critical failure mechanism, especially for surface mount technology, although there are several alternatives.

(2) In the *empirical approach*, no attempt is made to understand the process involved in failure. Large numbers of tests on actual boards are performed, usually under simulated and accelerated service conditions, or existing lifetime data are analysed and the *probability* of failure is assessed. It is assumed that the established relationships for failure times persist, and a statistical approach is adopted to assess reliability.

In comparison with other engineering applications, such as pressure vessels or gas turbine engines, electronics equipment is better suited to this treatment because of the smaller sizes and unit costs involved. However, as was demonstrated in Chapter 4, there are numerous variables to be considered – board (dimension, materials), component (type, size, number, arrangement), service conditions (temperature extremes, dwell times, heating and cooling rates) and solder alloy (composition and prior history). As a consequence, without any well-founded relationships between the key parameters, the uncertainty associated with each of them, and the effect on performance, this purely empirical approach can become time consuming and expensive. For example, thermomechanical fatigue involves initiation and growth of cracks until the net section stress levels are high enough to cause catastrophic failure. Three key parameters are likely to be the temperature range (equivalent to the stress or strain range), the initial defect size and the rate of crack growth. There may also be interactions between these parameters, and each has its own specific variability. It is not surprising, therefore, that in practice, a significant spread of failure times (or cycles) is observed. Of course, the true situation is more diffuse as the factors cited above should also be considered. In such circumstances, a probabilistic approach, involving numerous specimens tested under simulated service conditions, represents a sensible alternative. Consideration of the data on a probabilistic basis enables the global performance perspective to be addressed. Wymyslowski (1) has demonstrated that with a limited number of variables and no interaction between them, there is usually good agreement between the probabilistic and deterministic approaches. The former should always be reinforced by experimental evidence, however.

Electronics is a rapidly changing technology. Miniaturisation and the demands for higher power density involve the use of different (alternative) materials and the likelihood of changes in the critical failure mode. Both affect reliability. Design cycles need to be short and any accelerated testing should be representative and accurate to minimise costly prototyping and qualification testing. On the other hand, over-conservative design is inefficient.

The present chapter focuses on this empirical and statistical approach, with an emphasis upon actual results rather than the statistics themselves, although a few of the basic fundamentals are introduced [see refs (2-4) for more comprehensive treatments]. The mathematics involved is assumed knowledge, or can be obtained from any standard text. The major objectives of the chapter are to appraise the approach, to examine the extent to which acceleration can be achieved, and to demonstrate how the data it produces may be employed in the assurance of reliability.

2 FUNDAMENTALS

Reliability of a component, product or system, R, may be defined as the probability that it will perform its intended function for a specified period of time, under a given operating condition, without failure. In this sense, it is user- and application-specific.

Numerically, reliability can be described as the percent of survivors, and may vary between 0 and 1 (or 0–100 per cent) with total confidence set at the upper limit. Of course, reliability of a system may vary with time, (amongst other parameters), and the time-dependent reliability, R(t), is given by

$$R(t) = \frac{n_s(t)}{n_i} \tag{14.1}$$

where n_i is the total number of samples and n_s the number surviving after time, t.

2.1 The language of statistics

There are several terms that are in general use in the statistical treatment of data from electronics service or testing. These will be introduced by means of a simple example (5). If, say, one hundred measurements are taken of a property, the data may be represented in many forms. For example, a *histogram* shows the frequency or number of measurements (of resistance in this case) falling into a pre-selected interval or range of the property. The width of the block indicates the size of the range and the height denotes the number of measurements falling into that range (Figure 14-1).

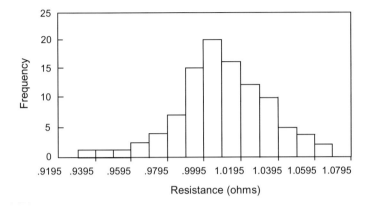

Figure 14-1. Histogram showing distribution of measurements (5)

A *frequency bar chart* employs bars centred on the mid-point of each range, with the height of the bar, again, signifying the frequency of occurrence of measurements in the range (Figure 14-2).

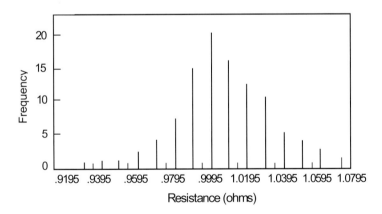

Figure 14-2. Frequency bar chart for the measurements shown in Figure 14-1

Note that 'frequency' could relate to any property, such as strength, cycles to failure or time to rupture. If the tips of the bars are joined a *frequency polygon* is produced which appears as a smooth curve when a large number of measurements is taken and the interval width (or range) is small (Figure 14-3). This curve represents the *probability*, or *frequency of occurrence*, and is a *probability density function* (PDF) of the property. The PDF, $f(x)$, is the probability of obtaining the value x for the property

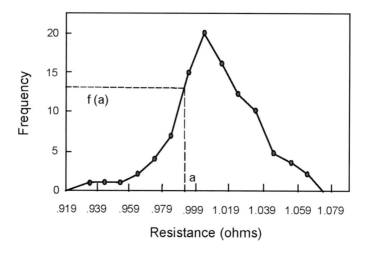

Figure 14-3. Frequency polygon for the resistance measurements (5)

$$f(x) \geq 0 \; ; \text{ and } \int_{-\infty}^{\infty} f(x)dx = 1 \qquad (14.2)$$

An alternative method is to consider the cumulative distribution of data, by computing the number of data points that fall below a class boundary. The *cumulative distribution function*, CDF, is $F(x)$, and has a value between 0 and 1. From Figure 14-4, the cumulative probability at a is given by

$$F(a) = \int_{-\infty}^{a} f(x)dx \qquad (14.3)$$

or $$f(x) = \frac{dF}{dx} \qquad (14.4)$$

and indicates that some 75 per cent of the readings were below 1.0250 ohms.
The *population median* is the value of x at which F(x) has a value of 0.5 (50 per cent).

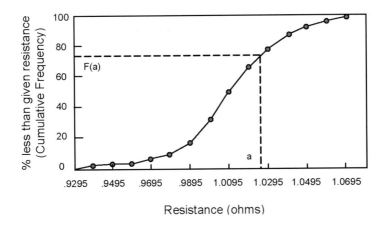

Figure 14-4. A cumulative distribution plot of the measurements shown in Figure14-1 (5)

The cumulative distribution function, F(x) is sometimes described as the *Life Distribution*, and may describe either the probability of a random unit in a population failing within x, or the fraction of all units in the population which fails within the same limit.

The CDF is related to reliability, R, as follows

$$R(x) = 1 - F(x)$$
(14.5)

and has the dual interpretation described above, i.e. R (x) is the probability of a random element still operating after x hours, or the fraction of units surviving at least x hours.

The *instantaneous failure rate*, (*or hazard rate*) h(x) represents the instantaneous rate of failure for units of a population that have survived to time, x. It is given by

$$h(x) = \frac{f(x)}{R(x)}$$
(14.6)

A unit used for failure rate is parts per million per thousand hours (ppm/K) or FIT (fails in time). For example, one ppm/K (one FIT) means that one failure is expected out of one million devices operating for one thousand hours.

The 'bathtub' profile of the hazard rate (Figure 14-5) is a popular method of representation of failure rate variation with time. An initial fall in rate is associated with the so-called 'infant mortality' as inherent defects, often introduced during manufacture, cause early failure. A period of stability or constant hazard rate follows when failures are due to random events. Finally, the gradual wear out processes (the failure mechanisms, such as crack initiation and growth, discussed elsewhere) dominate, and the hazard rate increases.

It is important to remember that life prediction approaches based upon the physics of failure strategy are concerned only with this final phase. From an assumption of an initial damage level (e.g. zero or a fixed dimension) they compute the time (or cycles) required to reach a critical ('mortal') defect dimension, using an expression for the rate of damage accumulation.

An important factor is the *scatter of life times* about the mean time to failure, since this denotes the confidence level that can be placed upon the median value i.e. the wider the scatter, the lower the confidence.

The *cumulative failure rate*, H(x), prior to x is defined as

$$H(x) = -\ln R(x) \tag{14.7}$$

and the *average failure rate*, AFR(x_1, x_2) between time, x_1 and x_2, ($x_2 > x_1$) is given by

$$\text{AFR}(x_1, x_2) = \frac{\ln R(x_1) - \ln R(x_2)}{x_2 - x_1} = \frac{H(x_1) - H(x_2)}{x_2 - x_1} \tag{14.8}$$

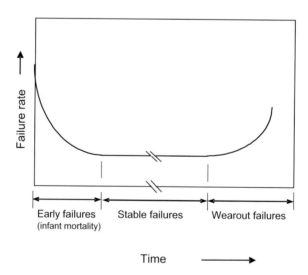

Figure 14-5. Failure rate as a function of time (bathtub curve

The *mean time to failure*, MTTF, is defined as the expected time to first failure, and may be calculated using the probability function, $f(x)$.

$$\text{MTTF} = \int_0^\infty xf(x)dx \tag{14.9}$$

Some workers (3) determine MTTF at the time at which $R(t) = 0.5$.

2.2 Example

The application and interrelation between these terms can be demonstrated with the following example (5). If the life distribution of a component is experimentally determined as

$$F(x) = 1 - (1 + 0.002x)^{-1}$$

Calculate
1. the probability that a new unit will fail by 4000h
2. the proportion surviving past 9000h
3. the failure rate at 10,000h
4. the average failure rate between 2000 and 9000h
        Solution – Recalling the relevant equations (14.4 -6, 14.8)

$$F(x) = 1 - (1 + 0.002x)^{-1}$$

$$R(x) = 1 - F(x) = (1 + 0.002x)^{-1}$$

$$f(x) = \frac{dF}{dx} = 0.002(1 + 0.002x)^{-2}$$

$$h(x) = \frac{f(x)}{R(x)} = 0.002(1 + 0.002x)^{-1}$$

Then
5. $F(4000) = 0.889$
6. $R(9000) = (1 + 0.002 \times 9000)^{-1} = 0.053$
7. $h(10,000) = 0.0001$ per hour
8. $AFR(2000,9000) = \dfrac{(\ln 0.2 - \ln 0.053)}{9000 - 2000} = 0.0002$ per hour

2.3 Life distributions

The life distribution, or cumulative distribution function, curve describing the spread of experimental data may take many forms, such as exponential, normal and log normal, although for electronics components, solder joint fatigue and wear out, the *Weibull life distribution* is more appropriate (Figure 14-6). It is a two-parameter (θ, β) function, and can accommodate situations in which the percentage occurrences decrease, increase or remain constant with increase in the characteristic measured. Using this life distribution, the Weibull expressions for PDF, CDF, $R(x)$ and $h(x)$ are given below

$$f(x) = (\beta / x)(x / \theta)^{\beta} \, e^{-(x/\theta)^{\beta}} \tag{14.10}$$

$$F(x) = 1 - e^{-(x/\theta)^{\beta}} \tag{14.11}$$

$$R(x) = 1 - F(x) = e^{-(x/\theta)^{\beta}} \tag{14.12}$$

$$h(x) = \frac{f(x)}{R(x)} = (\beta / \theta)(x / \theta)^{\beta-1} \tag{14.13}$$

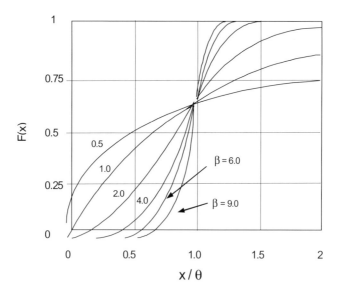

Figure 14-6. Weibull cumulative distribution function (CDF)

Clech et al (6) advocate the use of a three parameter Weibull distribution to include *failure free time*. This provides improved fit for the early stages of wearout. In conjunction with their Comprehensive Surface Mount Reliability Model (7) and using the Miner Cumulative Damage law, if appropriate, they provide several detailed examples of the efficacy of this approach. While this source is valuable for potential practitioners, the present chapter will remain with the two-parameter approach.

2.4 Application

A study of the endurance of wire bond interconnects (1) can be used both as a demonstration of the probabalistic approach and as the subject of scrutiny by a traditional engineer.

During thermal cycling, fatigue cracks initiate and grow until separation, and open circuit occurs. From Chapters 3 and 13, the fatigue life can be calculated by integrating the growth rate expression between an initial and final crack dimension (a_0 and a_f respectively) i.e.

$$N_f = \frac{1}{C} \int_{a_0}^{a_f} \frac{1}{(\Delta K)^m} da \qquad (14.14)$$

C and m are the constants in the crack growth rate equation. N_f was computed as 1.2×10^7 cycles assuming that $a_f \gg a_0$; the number of cycles involved in crack propagation, N_p, is much greater than those required for initiation, N_i; and the stress range, $\Delta\sigma$, was equal to the maximum stress, σ_{max}.

It was demonstrated that $\Delta\sigma$, C and a_0 affected the value of N_f with diminishing significance. Using these input parameters, together with several forms of probability distribution for fatigue data, values of the cumulative distribution function F (N_f) at 0.5 were derived. The coincidence between the deterministic value of N_f and the cumulative distribution function N_{f50} was very good, with five out of six values falling within ±25 per cent. The sixth, which assumed a stochastic interaction between the input variables was a factor of three smaller – not a startling difference where fatigue data are concerned.

Considering the assumptions stated above, the dissimilarity between initial and final crack lengths is acceptable (apart from defective structures that fail rapidly). The use of the maximum stress level for the stress range ignores crack closure at stresses above zero, which result in an 'effective' stress range smaller than that employed. Probably most significant is the assumption that $N_p \gg N_i$. A general rule of thumb is that for high strain fatigue, when the plastic strain component of the total strain exceeds the elastic element, the propagation phase does dominate. This occurs typically at lifetimes of around 10^4 cycles, but for greater endurances and predominantly elastic conditions, the number of cycles for crack initiation can constitute up to 90 per cent of the total life. The lifetimes in the reported study were around 10^6 to 10^7 cycles. This potentially large degree of conservatism is further augmented on closer inspection of the crack growth rate curve. The early stage growth rates are substantially lower than those expected from extrapolation of the Paris equation, $\dfrac{da}{dN} = C\Delta K^{m}$ (Equation 3.11). These reservations are generic, and fortunately conservative. The solder situation is further exacerbated by the existence of the intermetallic layer and an unstable microstructure, the effects of which on crack initiation and growth are generally unknown.

3   ACCELERATED TESTING AND STATISTICAL ANALYSIS

As discussed previously, reliability is usually measured by *accelerated testing,* in which key parameters, such as temperature, temperature range, stress and heating or cooling rates are accentuated relative to those experienced in service. To further reduce testing time, the extent of dwell periods in the operational profile is substantially reduced. Clearly, in a competitive field such as electronics, the time to market is extremely important. However, if the level of acceleration is too high, there is a danger that the processes causing failure in the laboratory test may differ from those that control performance in service. In this case, these empirical data are virtually worthless. The present section considers how data from accelerated test programmes may be accommodated in statistical analysis.

Engelmaier lists nine common use categories of electronic assemblies in increasing order of severity in the conditions they experience (3). Associated with these categories are the recommended test conditions for accelerated thermal cycling. It is notable that the minimum recommended dwell time is 15 minutes. When the service environment includes 'cold' and 'hot' elements, the accumulation of both time-dependent and time-independent damage processes is required. (It is assumed that only time-independent damage occurs at temperatures below –20°C, although in Chapter 5, it was demonstrated that time-dependent creep of solder alloys occurs at –50°C.)

The number of failure-free test cycles necessary to ensure that the failure risk per component is below a stated cumulative failure probability (1, 10 and 100 ppm) at the end of the stated service life, with 32 components on test, is also quoted. Increasing the number of components under test reduces the required number of cycles – but there is a limit to this strategy. For applications which demand low   failure risks, the number of failure-free cycles is prohibitively high and may exceed 10^{5} cycles. Difference between leadless and leaded-solder attachments, in terms of accelerated testing, are also apparent. Each type has a different Weibull modulus and a different sensitivity to temperature range. Figure 14-7 demonstrates this effect for one specific chip carrier in terms of the cumulative failure probability and operating time. The leadless attachments yield shorter lives and wider scatter.

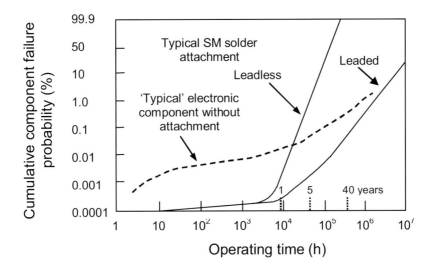

Figure 14-7. Cumulative failure probabilities for surface mount components illustrating difference between leaded and leadless attachments (3)

3.1 Factors for achieving acceleration

The prime target for reducing test times is the dwell period, which in service may extend to hours or even days. In contrast, dwells of a few minutes are common in accelerated thermal cycling tests, and sometimes as low as a few seconds in isothermal mechanical fatigue tests. It is essential that this restriction does not eliminate the damage processes that give rise to failure in service.

Enhancement of the applied strain range by increasing the temperature limits produces only modest accelerations and introduces the possibility of thermal shock. Strain range may also be increased by artificially enlarging the CTE mismatch by selecting other materials for the joint. Extensive advice regarding the factors for acceleration and their limitations has recently been published in an IPC Standard (8).

3.2 Types of acceleration test

Functional accelerated testing combines both external thermal cycling with internal power dissipation cycling, and is most representative of operational conditions, apart from dwell times.

Thermal cycling is simpler. The rates of heating and cooling should be less than 30°C per minute and the waveform should be trapezoidal.

Isothermal mechanical cycling produces comparative data between alloys, and may be benchmarked against low acceleration tests. There is usually no difference in failure mechanism during thermal and mechanical cycling.

In planning tests for statistical evaluation, it is recommended that a minimum of 32 daisy chains should be used. (A daisy-chain test net should have six or more solder joints.) There is evidence that defects such as misalignment of pad and lead, gas bubbles, shrinkage voids and small solder volume are not detrimental to long term reliability. Gross defects are revealed in the infant mortality phase.

To account for ageing and grain growth in service, prior ageing may be applied. The equivalent ageing time may be estimated from

$$\ln\left(\frac{t_2}{t_1}\right) = \frac{Q}{k}\left[\frac{1}{T_2} - \frac{1}{T_1}\right]$$
(14.15)

where Q is the activation energy (0.42 ev for grain growth in eutectic Sn-Pb solders) and k is the Boltzmann constant. Otherwise, fine grains lead to longer fatigue lives in accelerated testing. A pre-ageing treatment of 300h at 100°C has been suggested as appropriate to produce sufficient microstructural coarsening, intermetallic and grain growth. The effect of such a treatment and whether it is realistic have not been properly addressed, although differences in creep behaviour in the as-cast condition and after a stabilising treatment such as this, have been described in Chapter 5.

3.3 Incorporation of acceleration into statistical analysis

If a random variable, such as time to failure during service operation, is x_0 and its equivalent in accelerated testing is x_t, then the linear *acceleration factor*, AF, is x_0/x_t

Substituting for x_0, the principal statistical expressions may be written as:

$$f_0(x_0) = \frac{1}{AF}\, f_t\left(\frac{x_0}{AF}\right)$$
(14.16)

$$F_0(x_0) = F_t\left(\frac{x_0}{AF}\right)$$
(14.17)

$$h_0(x_0) = \frac{1}{AF}\, h_t\left(\frac{x_0}{AF}\right)$$
(14.18)

$$h_0(x_0) = \frac{1}{AF}\, h_t\left(\frac{x_0}{AF}\right)$$
(14.19)

where $f_0(x_0)$, $F_0(x_0)$, $R_0(x_0)$ and $h_0(x_0)$ are the PDF, CDF, reliability function and failure rate at the operating condition respectively.

With a Weibull distribution, the expressions become

$$F_0(x_0) = 1 - e^{-(x_0/AF\theta)^{\beta}} \qquad (14.20)$$

$$R_0(x_0) = e^{-(x_0/AF\,\theta)^{\beta}} \qquad (14.21)$$

$$h_0(x_0) = AF^{-\beta}(\beta/\theta)(x_0/\theta)^{\beta-1} \qquad (14.22)$$

When the dominant failure mechanism is a thermally activated process, such as creep, stress relaxation or corrosion, determination of the acceleration factor is straightforward *provided that the mechanism is the same in service and during testing.* If ΔQ is the activation energy of the dominant failure mechanism, then

$$AF = e^{(\Delta Q/k)(1/T_0 - 1/T_t)} \qquad (14.23)$$

k is Boltzmann's constant (8.62×10^{-5} ev/K). T_0 and T_t are respectively the operating and test temperatures in Kelvin.

As previously described, the principal factors governing thermomechanical fatigue of solder joints are temperature (maximum and range) and frequency (cycle time which incorporates dwell periods). Based upon the Coffin-Manson expression, values for the acceleration factor may be determined from one of the following equations,

$$AF = \left(\frac{f_0}{f_t}\right)^m \left(\frac{\Delta T_t}{\Delta T_0}\right)^n \frac{\phi_0}{\phi_t} \qquad (14.24)$$

$$AF = \left(\frac{f_0}{f_t}\right)^m \left(\frac{\Delta T_t}{\Delta T_0}\right)^n e^{1414\left(1/T_0\ 1/T_t\right)} \qquad (14.25)$$

$$AF = \left(\frac{f_0}{f_t}\right)^m \left(\frac{\Delta\gamma_t}{\Delta\gamma_0}\right)^n e^{1414\left(1/T_0 - 1/T_t\right)} \qquad (14.26)$$

f is the temperature cycling frequency, and as before, the suffixes 0 and t relate to the operating and test temperatures respectively; ΔT and $\Delta \delta$ refer to the temperature and shear strain ranges; ϕ is the isothermal fatigue life at maximum temperature within the cycle. The values of m and n are 0.33 and 1.9 – 2.0 respectively (9).

Equation 14.25 is the most convenient to use since no additional forms of data are required. The time to failure, t, is proportional to exp (Q/kT) and, if not known, the activation energy can be determined by measuring the mean time to failure at three temperatures, at least, and plotting against 1/T (K) on semi log paper. Equations 14.23 and 14.26 require data on isothermal fatigue performance or information from FE analysis regarding the shear strain range values.

3.4 Example(5)

A solder joint reliability thermal cycling test involves four boards (Figure 14-8) each containing 25 plastic quad flat packs (PQFP) with 208 pins attached to each (i.e. in total 100PQFPs and 20 800 solder joints). They were subjected to a thermal cycle (–40 to 125°C) each hour. The numbers of cycles to failure (defined as the first solder joint failure in any of the 208 pin PQFP) were analysed by the Weibull method and the best fit values obtained for β and θ were 3.66 and 6357 respectively. The performance under test is shown in Figure 14-9.

The service operating cycle is one per day ($f_0 = 1/24$ per hour) and the temperature ranges from 0 to 85°C. Data from the literature indicate that the ratio of the isothermal fatigue lives at 85 and 125°C is 1.7.

From equation 14.23, the acceleration factor may be determined

$$AF = \left(\frac{1}{24}\right)^{\frac{1}{3}} \left(\frac{165}{85}\right)^2 (1.7) = 2.2$$

The life distribution, reliability function and failure rates are then

$$F_0(x_0) = 1 - e^{-(x_0 / 14113)^{3.66}}$$

$$R_0(x_0) = e^{-(x_0 / 14113)^{3.66}}$$

$$h_0(x_0) = 0.0000311 \left(\frac{x_0}{6357}\right)^{2.66}$$

Figure 14-10 demonstrates that the one and fifty percent cumulative failure points are 11.8 and 34.7 years respectively, *assuming that no other damage mechanism influences behaviour.*

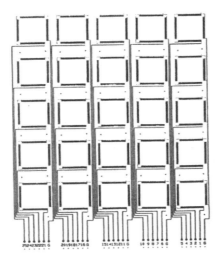

Figure 14-8. A test board for the 208-pin PQFP (5)

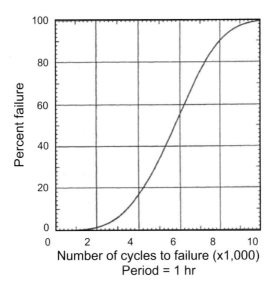

Figure 14-9. Life distribution of the 208-pin solder joints under test condition, -40 to 125°C, one cycle per hour (5)

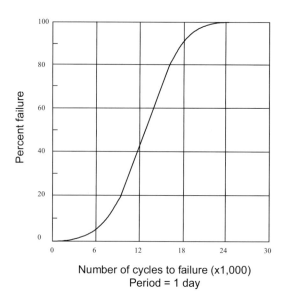

Figure 14-10. Life distribution of the 208-pin solder joints under operating condition, 0 to 85°C, one cycle per day (5)

4 CONCLUSIONS

The inherent variability that exists in materials, service conditions and board configurations means that a statistical analysis of performance data can make a valuable contribution to efficient design and timely production. While the approach is often conservative, it should be realised that not all influential parameters are considered. There is much scope for further exploration of this topic

Statistical analysis of accelerated test data makes a valuable contribution to the assurance of reliability when the degree of acceleration is a compromise between the conflicting demands of conservation of time (efficiency) and faithful simulation of the service condition. It is essential that the principal failure mechanism should be the same as that which dominates in service.

5 REFERENCES

1   A Wymyslowski, 'Probabalistic approach to numerical reliability assessment of microelectronic components' Proc. Second Int. Conf on Benefiting from Thermal and Mechanical Simulation in Microelectronics, (SIME 2001), Paris, 2001, 67-73.
2   J H Lau and Y H Pao, 'Solder Joint Reliability of BGA, CSP, Flip Clip, and Fine Pitch SMT Assemblies'. McGraw Hill, (New York) 1997.

3   W Engelmaier 'Solder attachment reliability, accelerated testing, and result evaluation' in 'Solder Joint Reliability : Theory and Applications', J H Lau (Ed), Van Nostrand, Reinhold, (New York) 1991 p545-587.

4   P M Stipan, B C Beihoff and M C Shaw, 'The Electronic Packaging Handbook', CRC Press (Boca Raton Florida), 2000, p15-1.

5   J H Lau and Y H Pao, 'Solder Joint Reliability of BGA, CSP, Flip Clip, and Fine Pitch SMT Assemblies'. McGraw Hill, (New York) 1997, Chapter 2.

6   J-P Clech, D M Noctor, J C Manock, G W Lynott and F E Bader, Proc. 44th ECTC (Washington DC), 1994, 487.

7   J-P Clech, J C Manock, D M Noctor, F E Bader and J A Augis, Proc 43rd ECTC (Orlando), 1993, 62.

8   Performance Test Methods and Qualification Requirements for Surface Mount Solder Attachments, IPC-9701, January, 2002.

9   H D Solomon, J Electronic Packaging, 1991, 113, 186.

CHAPTER 15

THERMAL AND MECHANICAL SIMULATION IN MICROELECTRONICS

1 INTRODUCTION

To even the most casual observer, it should appear obvious that microelectronics plays a vital role in modern human life and society. Since 1965, Moore's law, forecasting "the density of chip circuitry doubles every 18 to 24 months", has proved to be valid. There is tremendous pressure on companies to bring new and reliable products to market. However, the traditional method of design and prototyping may not be suitable for this new and highly competitive marketplace.

There are at least two main motivations to develop virtual thermo-mechanical prototyping of microelectronics:

First, thermo-mechanical reliability of microelectronics is a major concern for the electronics industry. Currently, about 65 per cent of all failures in microelectronics are related to thermal-mechanical problems (1). This is expected to increase due to further miniaturisation and function integration, which cause:

- an increase in power dissipation density
- a higher interconnection density and
- a demand for higher reliability

Secondly, most thermo-mechanical reliability problems originate from the product and/or process design phase. The trial-and-error method (designing, building and testing physical prototypes) for thermal-mechanical design still prevails, although in a climate of severe competition, legislation and technical change, these methods are no longer seen to be competitive, primarily because of the time they consume. There is an urgent need to develop innovative thermal-mechanical design and reliability qualification methods. While there are many strands of ongoing research into the building blocks for the virtual thermal-mechanical prototyping method, the two most vital are (1):

Reliable and efficient Finite Element Method (FEM)-based thermo-mechanical simulation models and advanced-simulation based optimisation methods. The latter includes optimisation components, e.g. maximum and minimum, parameter sensitivity and robust design.

The main advantage of using this type of virtual thermo-mechanical prototyping method is that the design requirements and reliability qualifications can be integrated, conducted and optimised at the earlier phase of the product design process. The results can be used to "predict, evaluate, qualify and eventually optimise the thermal and mechanical behaviour of microelectronic products against the actual product requirements, prior to major physical prototyping and manufacture" (1).

In other words simulation offers many benefits, if used correctly. These would include: (2)

- Optimised product performance and cost
- A reduction of development time
- An elimination or reduction of testing
- First time achievement of required quality
- Improved safety
- Satisfaction of design codes
- Improved information for engineering decision making
- Fuller understanding of components allowing more rational design
- Satisfaction of legal and contractual requirements

This explains the massive increase in the use of simulation to expedite the design process. There are however, many forms of simulation. The appropriate form is dictated by the physics of the problem and what questions the investigator is asking. To place the modelling issues into perspective, first consider the nature of models, as used within the context of Science and Engineering.

2 EXPERIMENTAL AND THEORETICAL MODELS

Modelling is a basic procedure or tool used to investigate properties of interest in some real phenomenon or system. Models may be experimental or theoretical.

Experimental models are usually of the same physical nature as the actual phenomenon. These experimental models reproduce the phenomenon in a simplified form, and sometimes, at a different scale.

Theoretical models on the other hand, represent the real phenomenon by using abstract concepts and usually employ mathematical methods of analysis. A theoretical model can be numerical (computational) or analytical (closed form solution, exact mathematical solution). Within Engineering, the most widely used numerical modelling tool is Finite Element Analysis (FEA). The ultimate goal of any theoretical modelling is to illuminate relationships, which while they exist, are obscured in the initial information. That is to say, the theoretical model can only provide information contained within the original data set, while experimental models could lead to new information.

Although an experimental approach, unsupported by theory, is 'blind', a theory, not supported by an experiment, is 'dead'. It is the experiment that forms a basis for a theoretical model, provides the input data for theoretical modelling, and determines the viability, accuracy, and limits of application of a theoretical model." (3)   In other words, theoretical modelling (e.g. numerical simulation) does not replace experimental modelling (testing). Numerical simulation should augment and enhances any testing.

2.1 Experimental Modelling

Experimental and theoretical models have their strengths, weaknesses and areas of proper application. Both are indispensable in the design of a viable, reliable, and cost-effective product.   While the role of theoretical modelling has increased significantly in other areas of engineering, the majority of studies dealing with mechanical behaviour and performance of microelectronics materials, components and devices, are experimental. (3)   There are several reasons for this.

- Experimental modelling can be carried out without theoretical support i.e., testing alone can be used for proof of viability and reliability of a material or product.
- Experiments involving microelectronics are substantially less expensive than in other areas, e.g. civil engineering or aerospace. Full-scale tests in these areas are prohibitively expensive and difficult to conduct. Experiments in microelectronics are much easier to design and conduct than in other areas of engineering, and this partly explains why the development of theoretical models in microelectronic applications has lagged behind.
- In microelectronics, materials whose properties are unknown are often successfully employed, e.g. thin film systems. This lack of knowledge of material properties is often an obstacle to implementing theoretical models.
- Many of the leading investigators in microelectronics traditionally use experimental methods as their primary tool and do not feel the addition of theoretical modelling will make a significant difference. This belief may no longer be viable.

It should be clear from the above that experimental modelling, although a requirement in certain areas, involves significant amounts of time and may incur considerable expense. Testing alone may be inadequate for understanding the mechanical, electrical or optical behaviour of a material or structure. This is because experimental data usually reflect the combination of many factors, which affect any given phenomenon. What is required for the prediction of behaviour is an in-depth understanding of the role of each parameter. This lack of insight usually leads to laborious, time-consuming and costly trial-and-error experimental procedures. Further, these procedures cannot, in general, be extended to new designs that are different from those currently being tested. Thus new experimental procedures have to be initiated for each new product. In a highly competitive environment, such as microelectronics, where time to market is a critical factor, it is self-evident that a better approach is required.

2.2 Theoretical Modelling

The ideal theoretical model produces easy-to use relationships and clearly indicates the role of the main factors affecting a given phenomenon. However, what are the benefits of theoretical modelling? If one is going to employ theoretical models, there should be advantages over a purely experimental approach.

Theoretical modelling should be able to predict the results of an experiment in less time and at a lower cost. It should also act as a guide in identifying valuable or worthless experiments. This information gives the investigator an opportunity to decide, how and what should be tested or measured, and in what direction success might be expected. It also establishes benchmarks for any subsequent numerical analysis or experiments by indicating the factors that affect certain outcomes.

Theory also forms a bridge between different experiments by providing a platform for interpreting different experimental results.

The complexities involved in taking into account all of the parameters that are relevant to a given physical system force the investigator to define some form of *equivalent system* with only those parameters that are essential and actually characteristic of the physical system under study. These parameters are known as *state variables*. The first step in the mathematical modelling of a physical system is the mathematical definition of the equivalent system. The graphical representation, of the process is illustrated in Figure 15-1 (4),

In general, with respect to computational mechanics, a set of differential equations is used to represent the equivalent system. These equations are derived through the use of fundamental principles of continuum mechanics.

The continuum models have an infinite number of degrees of freedom or parameters, since the *field variable*, e.g. temperature, is continuously distributed across the total domain of the given problem. An exact mathematical solution to continuum problems is generally impossible. While research continues in these formal or analytical methods, most attention dealing with 'real-life' problems is directed towards the formulation and application of discrete models which, by their very nature, are approximations.

Discrete models are obtained by considering additional simplifying assumptions. When these simplifications are introduced into a continuous model, the field variable can then be expressed through a finite number of parameters or degrees of freedom of the model. The process of converting from a continuous system to a discrete system is illustrated in Figure 15-2, and is known as the *discretization process*. The methods used to obtain a solution to these discrete models use the tools of numerical analysis. For example, Finite Element Analysis is one of the most popular and successful numerical analysis tools available to engineers for the solution of problems associated with microelectronics applications. For a more in-depth treatment of this topic the interested reader is referred to (4).

At this point a note of caution is appropriate to avoid the pitfalls that many have experienced with FEA. In certain parts of the Engineering community, there is often an illusion of simplicity in applying finite-element procedures. Some users of commercial finite-element code believe that they require no prior knowledge of structural analysis and materials engineering, and that the 'black box code' they deal with will automatically provide the right answer via a 'pretty picture' of some stress, strain, or temperature plot. A hasty, ill-considered, and incompetent application of computational analysis tools can result in more harm that good, by giving the impression that a real solution was obtained when, actually, this 'solution' may be simply wrong. It is usually quite easy to obtain a finite-element solution but it might be very difficult to obtain a correct and accurate solution. One has to have a good background in the engineering and physics of the problem, as well as an understanding of the behaviour of the material, in order to develop an adequate, feasible, and economic model, and to correctly interpret the output information (3).

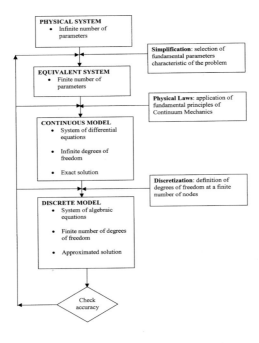

Figure 15-1 Mathematical-modelling process of physical systems

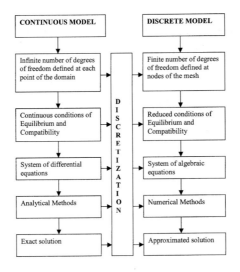

Figure 15- 2 Features of the discretization process

2.3 Analytical versus numerical (computational) modelling

Finite element analysis has been a powerful tool in the theoretical investigations of mechanical and structural engineering since the 1950s. This is not to say, however, that analytical procedures are unnecessary or less important. Clearly, analytical solutions to these problems have many advantages, not only for their clarity, but also by indicating the role of various factors affecting the behaviour of the system under investigation (3).

The development and use of computational models provides a way of understanding solder material behaviour that would be impossible or far too expensive to achieve through experimentation alone. Computational models also provide the ability to reduce the overall cost of designing and manufacturing new products (5).

3   FINITE ELEMENT ANALYSIS

The many possible forms of thermal and/or mechanical failure associated with microelectronic products, have been considered in earlier chapters. Clearly, there are substantial opportunities for failure to occur. To explore the role of simulation in determining the reliability of a given component, the Finite Element Method, as a numerical simulation tool, is now considered.

3.1 The Basics

Finite Element Analysis (FEA) is one of many numerical tools available to solve, what is in essence, a partial differential equation. It is up to the user to select the appropriate tools for the problem at hand. Although FEA is the predominant numerical analysis tool in modern engineering, the results obtained should not necessarily be considered as 'the answer'. This will be explored in later sections, but it is important to remember, that FEA is 'just a tool' and that the proper use of this tool is strictly the user's responsibility.

Because of the mathematical foundations of the finite element method, it is applicable to many classes of problems. Some of the more common application areas include: structural, fluid, dynamic, thermal and electromagnetic systems. Many of these are, of course, applicable to the problems encountered in microelectronics. In the following section some of features of general-purpose FEA systems are examined in more detail.

Although the fundamentals of the Finite Element Method are considered in Chapter 16, a few elementary definitions may be helpful at this stage.

Consider the following: a given body in which a distribution of some form of unknown parameter, say for example, displacement or temperature, is required. Prior to modelling, information with respect to the geometry of the body, its material parameters and initial, e.g. loads or temperatures, or boundary conditions, e.g. restraints or fixed temperatures, must be known or determined.

Begin by dividing the body into subdivisions know as *elements*. This process is known as *discretization* (the formal process was illustrated in Figure 15-2). These elements are connected at the joints, which are known as *nodes*. In the Finite Element Analysis process, it is assumed that the variable of interest is acting over every element in a predefined manner. The distribution of the variable across the element is defined in terms of some form of function, e.g. polynomial. The specification of the number, type and placement of these elements is the user's responsibility and these are chosen to approximate the problem at hand.

After the body has been discretized (split into elements), the governing equations for each element are calculated and are then assembled into a global system of equations, which are then solved to produce the desired, and hopefully the correct and accurate distribution of the variable, through each element and thus the entire body.

In the real world of modelling microelectronics, problems will arise. Typical ones likely to be encountered in FEA simulations would include:

- Incorrect or unavailable material parameters
- Missing knowledge of the processes at the connection areas of different materials, e.g. connecting a soft area to a hard area
- Missing material parameters of these non-bulk materials at the interfaces, e.g. thin layers, IMC
- Missing parameters for the countless variations of plastics
- Missing descriptions and equations for the manufacturing and environmental influences

While the above list is not exhaustive, it does point out the importance and necessity of proper and accurate material data. Unfortunately, obtaining this information is, in some cases, not an easy task.

3.2 Capabilities of Finite Element Programs

Finite element codes or programs fall within two main groups. (i) general-purpose systems with large element libraries, sophisticated modelling capabilities and a range of analysis types. (ii) specialised systems for particular applications, e.g. air flow around/over electronic components.

To provide a feeling for the expected capabilities of good general-purpose FEA system, some properties under the various major forms of analysis are considered. While systems usually offer many analysis areas, the most relevant to microelectronics are structural and thermal, with other areas being electro-magnetics and fluids. Many of these areas of analysis are capable of being coupled. For example, a common form of coupled analysis in microelectronics is thermal-stress analysis, where the results of a thermal load case are transferred to a stress analysis.

Some general capabilities of Finite Element Analysis codes, derived from a NAFEMS booklet (2), are summarized in Tables 15.1 through 15.4

Table 15-1 Linear Structural capabilities

Structural Linear
homogeneous/non-homogeneous materials
isotropic/orthotropic/anisotropic materials
temperature dependent material properties
spring supports
support displacements point, line, pressure loads
body forces (accelerations)
initial strains (e.g. concrete pre-stressing tendon)
expansion
fracture mechanics
stress stiffening
natural frequencies and modes of vibration
response to harmonic loading
general dynamic loading
response spectrum loading
power spectral density loading
spin softening
eigenvalue buckling

Table 15-2 Non-Linear Structural capabilities

Structural Non-Linear
material non-linearities (e.g. plasticity, creep)
large strain (gross changes in structure shape)
large displacements
gaps (compression only interfaces)
cables (tension only members)
friction
metal forming
time history response on non-linear systems
large damping effects
impact with plastic deformation

Table 15-3 Linear Thermal capabilities

Thermal Linear
Homogeneous / non-homogeneous materials
Isotropic / orthotropic /anisotropic materials
temperature dependent material properties
conduction
isothermal boundaries
convection
heat fluxes
internal heat generation
Time- dependent boundary condition

Table 15-4 Linear Thermal capabilities

Thermal Non-Linear
radiation
phase change

3.3 Results of Finite Element Analyses

The amount of information that can be produced by a FEA system, especially for non-linear analysis, is enormous, and to the first-time user can be daunting. Most general-purpose Finite Element codes provide the capability to determine many items from a given analysis. In addition, graphical presentation of selected items is also available. Tables 15.5 through 15.8 illustrate some of the generated results from various forms of analysis (2):

Table 15-5 Typical Information generated by a stress analysis

Stress analysis
Deflections
Reactions at supports
Stress components
Principal stresses
Equivalent stresses (Tresca, von Mises, etc)
Strain
Strain energy
Path integrals and stress intensity for fracture mechanics
Linearised stresses
Buckling loads
Buckling mode shapes

Table 15-6 Typical Information generated by a dynamic analysis

Dynamic analysis
Natural frequencies
Natural mode shapes
Phase angles
Participation factors
Dynamic analysis
Response to loading
Displacements
Velocities
Accelerations
Reactions
Stresses
Strains

Table 15-7 Typical Information generated by a Thermal analysis

Heat transfer
Temperatures
Heat fluxes

Table 15-8 General Information generated by an analysis

General
Displaced shape plots
Symbols showing the magnitude of reaction forces, heat fluxes, etc.
Contour plots of stresses, strains, displacements, temperatures, etc.
Vector plots showing the direction and magnitude of principal stresses, etc

The above lists are not exhaustive, but illustrate the quantity and type of information available to the analyst whose responsibility it is to understand the nature of the problem to be solved and the input and output data required. In the absence of such awareness, the system degenerates into a 'black box' category, and the solution it provides will almost certainly be inaccurate, if not wrong. It cannot be emphasised strongly enough that while most FEA systems produce vast amounts of data and pretty, highly persuasive pictures, it is the user's responsibility to ensure correctness and accuracy.

4 SUMMARY

Modelling is an important part of modern engineering and in particular, microelectronics. In the first part of the chapter, the nature of both theoretical and experimental models and their importance in solving real-life engineering problems was shown. Some of the advantages of using simulation tools were:

- Optimised product performance and cost
- A reduction of development time
- An elimination or reduction of testing
- First time achievement of required quality
- Improved safety
- Satisfaction of design codes
- Improved information for engineering decision making
- Fuller understanding of components allowing more rational design
- Satisfaction of legal and contractual requirements

With respect to the development of models, the most important, as far as Finite Element Analysis is concerned, is the discretization process, as represented by Figure 15-2. That is to say, the conversion of a real-life continuum problem into a discrete model for analysis, with the attendant assumptions, is critical in obtaining a model capable of giving both correct and accurate results.

FEA is a powerful tool for evaluating a design and making comparisons between various alternatives. It is not the universal panacea that replaces testing and allows users to design products without a through understanding of the engineering principles involved.

The qualification of assumptions is the key to the successful use of FEA in any product design. To achieve this, it is essential to:

- Appreciate the Physics and Engineering inherent in the problem
- Understand the mechanics of the materials being modelled
- Be aware of the failure modes the products might encounter
- Consider the manufacturing and operating environment of the product and how these might impinge on the performance
- Assume that the FEA results are incorrect until they can be verified
- Pay close attention to boundary conditions, loads and material models

The capabilities and the limitations of each element type selected, each solution method employed, and every boundary condition or load applied should also be appreciated. There is an assumption behind every decision, both implicit and explicit, that is made in finite element modelling.

The consideration of thermal and mechanical issues needs to become an integral part of product design and improved data, relating to thermal and mechanical properties of electronics materials, is paramount to good simulation. There remains, however an urgent requirement to develop more comprehensive models and to correlate those models with physical measurements.

5 REFERENCES

1   GQ Zhang, P Maessen, J Bisschop, J Janssen, F Kuper, *Virtual Thermo-Mechanical Prototyping of Microelectronics: The challenges for mechanics professionals,* in "Benefiting from Thermal and Mechanical Simulation in (Micro)-Electronics", EuroSimE2001, 2001, 21.
2   D Baguley, DR Hose, *Why do – Finite Element Analysis?,* NAFEMS, 1994
3   E Suhir, *Mechanical Simulation in Microelectronics and Photonics Packaging: Its Role, Merits, Shortcomings and Interaction of Experimental Techniques*, in "Benefiting from Thermal and Mechanical Simulation in (Micro)-Electronics", EuroSimE2001, 2001, 3.
4   A Portela, A Charafi, *Finite Elements Using Maple: A symbolic Programming Approach*, (Springer, 2002)
5   C Bailey, Editorial, Soldering & Surface Mount Technology: Computational Modelling, 2002, 14, No. 1, 2002

CHAPTER 16

FINITE ELEMENT ANALYSIS

1 INTRODUCTION

This chapter considers the basic theory and principles of the finite element method. The mathematics involved is described only briefly, as this is readily available in numerous textbooks. It is, however, important to understand the underlying mathematics to avoid the "black box" syndrome, which is likely to yield answers without any physical foundation.

1.1 The Solution of Boundary Value Problems

Given any physical problem that is governed by a differential equation, e.g. heat conducting through a body, the best way to solve it is to obtain the analytical solution. Unfortunately, in real-life engineering problems this is very difficult, if not impossible. For example, the region or the materials under consideration may be so complex that it is mathematically difficult to describe them, e.g. anisotropic materials or when there are non-linear terms in the equations which describe their behaviour.

In these situations, numerical methods can be used to obtain approximate solutions. Numerical solutions produce values at discrete points in space and/or time for one set of independent parameters. If these parameters change, then the solution procedure must be repeated. Even with this shortcoming, numerical solutions are more desirable than no solution! The calculated values provide important information about the physical process even though they are at discrete points

There are several methods for obtaining a numerical solution to a differential equation. While they will not be covered in detail, it is worth being aware of them. There are three general methods:

Finite Difference Method

The finite difference method approximates the derivatives in the governing differential equation using difference equations. This method works well in problems involving heat transfer and/or fluid mechanics. It is applicable to 2-D problems where the boundaries of the physical system are parallel to the coordinate axes. The major drawback is that the method is cumbersome when regions of the system under investigation have curved or irregular boundaries.

Variational Method

The variational method involves the integral of a function that produces a number. Each new function produces a new number. The function that produces the lowest number has the additional property of satisfying a specific differential equation. The process can be reversed. Given a differential equation, an approximate solution can be obtained by substituting different trail functions into the appropriate functional, i.e. a function of a function. The trial function that gives the minimum value is the approximate solution. The methods major drawback is that it is not applicable to any differential equation containing a first derivative term.

Weighted Residual Methods

The weighted residual methods also involve an integral. In these methods, an approximate solution is substituted into the differential equation. Since the approximate solution does not satisfy the equation, a residual or error term results. Various forms of this method exist including: Collocation, Subdomain, Galerkin's, and Least Squares. Of these, one of the most commonly used in FEA is Galerkin's method.

Integral methods have their strong and weak points. The important aspect here is that the numerical solution of a differential equation can be formulated in terms of an integral. The integral formulation is a basic characteristic of the FEM.

Another method included for completeness is the Direct Method, which is applicable to simple trusses, stepped bars and frames. The method is very difficult to apply to two- and three- dimensional problems, and so is limited to simple one-dimensional cases. For example, the solution to the simple stepped bar problem described later in this chapter uses this method. It is cited here as an illustration of the FEA method and because all parameters employed have a physical interpretation. In this method, the displacement caused by the applied forces is expressed by a set of equations convertible into a stiffness matrix for each of the structural members. These individual element stiffness matrices are assembled into a single stiffness matrix representing the entire structure. By applying appropriate boundary conditions and loads, the system of equations could be solved, by for example, back substitution.

2. GENERAL THEORY

The basic principles underlying the Finite Element Method are relatively simple (1). Consider a body in which the distribution of an unknown variable (displacement, temperature, etc.) is required. The region or body is subdivided into an assembly of subdivisions called *elements*. The elements are interconnected at joints, known as *nodes*. The complete set of elements is known as a *mesh*. A partial mesh is illustrated in Figure 16-1 and was also discussed in Chapter 15. Note that this is not a very good mesh in that (1) the triangular elements are of different sizes and shapes, and (2) it is not generally a good idea to have a single element, e.g. the large triangle(s) span the object.

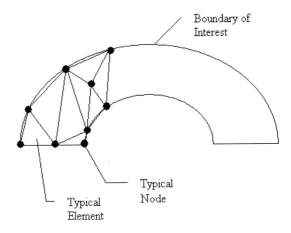

Boundary of
Interest

Typical
Node

Typical
Element

Figure 16-1 Example of a mesh over a given region

The field variable, e.g. temperature, is assumed to act over each element in a predefined manner. For example, the variable may assume a constant, linear or quadratic distribution. The distribution across each element is typically defined by a polynomial or trigonometric function. The number and type of element chosen must be such that the variable distribution through the whole body is adequately approximated by the combined elemental representations. For example, if the mesh is too coarse, the resolution of the parametric distribution may be inadequate, whereas too fine a mesh is wasteful of computing time and possibly the user's time in meshing.

After problem discretisation, i.e. subdividing the problem domain into discrete elements (meshing), the governing equations for each element are calculated and then assembled to give system equations. Once the general format of the equations of an element type, e.g. linear, is derived, the calculation of the equations for each occurrence of that element in the body is straightforward. Nodal coordinates, material properties and loading conditions of the element are simply substituted into the general format. The individual element equations are assembled into the system equations, which describe the behaviour of the body as a whole. These generally take the form:

$$[K]\{U\} = \{F\} \tag{16.1}$$

Where, in structural problems, $[K]$ is a square matrix, known as the *stiffness* matrix, $\{U\}$ is the vector of unknown nodal displacements (or temperatures); and, $\{F\}$ is the vector of applied nodal forces (or heat flux). The above equation is directly comparable to the equilibrium or load-displacement relationship for a simple one-dimensional spring, where a force, f produces a deflection, u in a spring of stiffness, k. To find the displacement developed by a given force, the relationship is inverted, i.e. u = k/f. The same approach applies to the FEM. However, before the equation can be inverted and solved for $\{U\}$, some form of boundary condition must be applied. In structural problems, the body must be restrained from rigid body motion. For thermal problems, the temperature must be defined at one or more nodes. The solution to the equation is not trivial in practice because the number of equations involved, tends to be very large. It is not unreasonable to have 250,000 equations, and consequently [K] cannot be simply inverted! Fortunately, [K] may be banded, i.e. terms are grouped about the main diagonal of the matrix, and techniques have been developed to store and solve the equations efficiently. After solving for the unknown nodal values, it is then a simple matter to use the displacements to find the strains and then the elemental stresses.

It must be emphasised that only the very basic concepts and mathematics of the Finite Element Method are presented here. For a more comprehensive coverage of the theory, the reader is referred to one of many excellent textbooks on the subject, for example, (2) and (3).

2.1. Stages of Model Creation

The major stages in the creation of any finite element model for most types of analysis are (4):
- Selection of analysis type
- Idealization of material properties
- Creation of the model geometry
- Application of supports or constraints
- Application of loads
- Solution optimisation

It is extremely important to:
- Develop a feel for the behaviour of the structure
- Assess the sensitivity of the results to approximations of the various types of data
- Develop an overall strategy for the creation of the model
- Compare the expected behaviour of the idealized structure with the expected behaviour of the real structure

3. STRESS ANALYSIS OF AN AXIALLY LOADED STEPPED BAR

To illustrate the process described above a simple example of an axially loaded stepped bar, Figure 16- 2a, using the direct method will be examined. First the finite element idealization is defined using only two one-dimensional elements (Figure 2b), although more elements could be used if desired. Each region is represented by a single element. That is, region (1) is represented by element (1) with nodes 1 and 2. Region (2) is represented by element (2) with nodes 2 and 3. Note that there is a common node, i.e. 2, shared by elements (1) and (2). This represents the interface between regions (1) and (2).

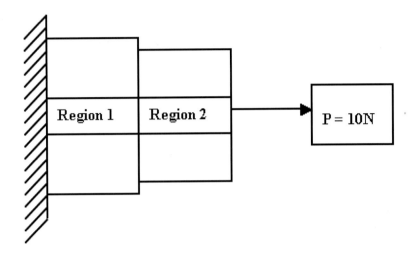

Figure 16-2a A simple stepped bar

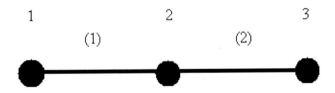

Figure 16- 2b The FEA model for the stepped bar

As the elements are one-dimensional, the nodes have only one degree of freedom (DOF), i.e. translation in the x-direction. In this case, we have two elements with a node at each end. This results in a total of three nodes and three DOF's.

The second step involves defining the material properties of the bar and the element connectivity. That is, what elements are connected to a given element. In this example,

$$A^{(1)} = 20mm^2, A^{(2)} = 10mm^2$$

$$L^{(1)} = L^{(2)} = 100mm$$

$$E^{(1)} = E^{(2)} = 200 \times 10^3 \ MPa$$

where A is the cross-sectional area, E is the Elastic Modulus, and L is the length of the regions.

Connectivity is very important because it contains details of how the elements are connected together, and how their stiffness matrices should be assembled into the global or system stiffness matrix. In our case, Element 1 has nodes labelled 1 and 2 and Element 2 has nodes labelled 2 and 3. From the mechanics of materials, k, the stiffness term, is given by AE/L and the displacement of a rod is given by u = FL/AE. The row and column indices of the matrices represent the relevant DOFs.

For element (1), the stiffness matrix is given by

$$\left[k^{(1)} \right] = \frac{A^{(1)} \times E^{(1)}}{L^{(1)}} \begin{bmatrix} 1 & -1 \\ -1 & 1 \end{bmatrix} = \begin{bmatrix} 4 & -4 \\ -4 & 4 \end{bmatrix} \times 10^4 \, N \, / \, mm \qquad (16.2)$$

and for element (2),

$$\left[k^{(2)} \right] = \frac{A^{(2)} \times E^{(2)}}{L^{(2)}} \begin{bmatrix} 1 & -1 \\ -1 & 1 \end{bmatrix} = \begin{bmatrix} 2 & -2 \\ -2 & 2 \end{bmatrix} \times 10^4 \, N \, / \, mm \qquad (16.3)$$

In the next step, assembly of the system stiffness matrix must be performed, where each element is added into the system matrix in turn, as determined by the connectivity. Element (1) has nodes 1 and 2, and its rows and columns are associated with the displacements u_1 and u_2. The row DOF positions are indicated in the following expression and are for illustration purposes only. They do not explicitly appear in the stiffness matrix. The column positions are similar, but of course run vertically.

$$\left[k^{(1)} \right] = \begin{matrix} u_1 & u_2 \\ \begin{bmatrix} 4 & -4 \\ -4 & 4 \end{bmatrix} \end{matrix} \times 10^4 \, N \, / \, mm \qquad (16.4)$$

Similarly, element (2) is associated with node 2 and 3, and its stiffness matrix can be labelled as

$$\left[k^{(2)} \right] = \begin{matrix} u_2 & u_3 \\ \begin{bmatrix} 2 & -2 \\ -2 & 2 \end{bmatrix} \end{matrix} \times 10^4 \, N \, / \, mm \qquad (16.5)$$

The system matrix, which has a row and a column associated with each degree of freedom, is therefore of size 3x3. It is filled by placing the terms from each *element stiffness matrix* into the correct location of the system matrix.

$$[k] = \begin{bmatrix} u_1 & u_2 & u_3 \\ 4 & -4 & 0 \\ -4 & 4+2 & -2 \\ 0 & -2 & 2 \end{bmatrix} \times 10^4 \, N \, / \, mm \tag{16.6}$$

No terms appear in the (u_1, u_3) locations, because no element connects these two degrees of freedom together.

The global force vector is the vector of applied nodal forces, that is

$$\{F\} = \begin{bmatrix} 0 & 0 & 10 \end{bmatrix} \tag{16.7}$$

since a force (of 10 N) is only applied at node 3.

The final system of equations is therefore

$$10^4 \begin{bmatrix} 4 & -4 & 0 \\ -4 & 6 & -2 \\ 0 & -2 & 2 \end{bmatrix} \begin{Bmatrix} u_1 \\ u_2 \\ u_3 \end{Bmatrix} = \begin{Bmatrix} 0 \\ 0 \\ 10 \end{Bmatrix} \tag{16.8}$$

Now constraint conditions must be applied to these equations before solving for the nodal displacements. In this case, $u_1 = 0$ and the equations are modified as

$$10^4 \begin{bmatrix} 4 & -4 & 0 \\ -4 & 6 & -2 \\ 0 & -2 & 2 \end{bmatrix} \begin{Bmatrix} 0 \\ u_2 \\ u_3 \end{Bmatrix} = \begin{Bmatrix} 0+R_1 \\ 0 \\ 10 \end{Bmatrix} \tag{16.9}$$

where R_1 in the first equation is the reaction provided by the surroundings at node 1 to maintain the zero displacement. Solution of the other two equations yields

$$u_2 = 0.25 \times 10^{-3} \, mm$$
$$u_3 = 0.75 \times 10^{-3} \, mm \tag{16.10}$$

and substitution back into the first equation above, gives $R_1 = -10N$ as expected. It is negative as it acts in the negative x direction.

Next, the strains, $\varepsilon^{(1)}$ and $\varepsilon^{(2)}$ in each element are calculated.

$$\varepsilon^{(1)} = \left(-u_1 + u_2\right)/L = 0.25 \times 10^{-3}/100 = 2.5 \times 10^{-6}$$

$$\varepsilon^{(2)} = \left(-u_2 + u_3\right)/L = 5.0 \times 10^{-6}$$

$$(16.11)$$

Finally, the stresses are found by application of Hooke's law:

$$\sigma^{(1)} = E\varepsilon^{(1)} = 0.5 N/mm^2$$

$$\sigma^{(2)} = E\varepsilon^{(2)} = 1.0 N/mm^2$$

$$(16.12)$$

The theoretical stresses for this problem are easily calculated and compared to the finite element results.

$$\sigma_{theory}^{(1)} = P/A^{(1)} = 10/20 = 0.5 N/mm^2$$

$$\sigma_{theory}^{(2)} = P/A^{(2)} = 10/10 = 1.0 N/mm^2$$

$$(16.13)$$

In this simple example, the finite element analysis has predicted the exact solution to the stepped bar. This is because the variation of displacement through each element was assumed to be linear, and this is indeed the case.

By selecting a linear distribution of displacement, a constant strain and therefore a constant stress field are imposed in each element, which is precisely what is expected in an axially loaded bar of constant cross-sectional area.

For simple elements such as rods or bars, the derivation of the elemental stiffness matrices is straightforward. However, for more advanced elements, this derivation is more complex and is usually makes use of some form of energy theorem.

4. PLANNING A FINITE ELEMENT ANALYSIS

Good planning is essential and to neglect it can render even the most elementary analysis both inefficient and incorrect.

4.1 The Purpose of the Analysis

The first step of the planning process entails the specification of the purpose of the analysis. Precisely, what is the objective of the exercise? The formal specification of this purpose at the initial stage of analysis has a bearing on the source of the data, the method of idealisation, the results to be produced, the accuracy required, the checking and validation required, and the selection of software and the allocation of staff. (5)

As this objective may not necessarily be straightforward, suitable time should be given to answering this question prior to attempting any form of analysis.

4.2 Accuracy (Not the same as Correctness)

All correct analyses i.e. appropriate geometry, mesh, loads and boundary conditions, may require some refinement to achieve the required level of accuracy. Accuracy is not the same as correctness. If a solution is correct, then accuracy is an issue. If the solution is incorrect, then accuracy is irrelevant and no further time should be expended in attempting accuracy. It is essential to obtain a verifiably correct solution first, then address the accuracy issues. Assuming a correct solution, then the accuracy of that solution can be affected by many factors, such as the assumption of linearity, the representation of adjoining structures, the material properties, the accuracy of the geometric representation, the loading, the mesh density, the element type, the shape types, and the numerical error in the solution (5).

4.3 Validation

Analysis validation, i.e. confirmation of methods, procedures, data and assumptions, is a very important aspect of any finite element analysis. It should involve, at a minimum, the following (5):
- Data checking, e.g. geometry, material, boundary conditions, to ensure that the model analysed is sufficiently close to that intended
- Checking that the required accuracy has been achieved
- Qualification of the assumptions made in the idealization process

The appropriate level of validation is a function of the end use of the results. The two extremes might be a purely investigative analysis with no inherent risk to life and limb or an analysis that is the only source of qualification, for a high risk, potentially catastrophic event if the structure or system fails.

In any case, the onus is on the user to carry out the necessary validation procedures that are appropriate to the analysis at hand. There are several basic checks that the user should always carry out on the output to confirm that the model is reasonable. According to (1) these would include:
- Volume and mass checks: to confirm that the geometry is represented correctly.
- Reaction force checks: the sum of all the reaction forces should equal the applied forces.
- Model equilibrium: the sum of the forces and resultant moments acting on the problem should equal zero.
- The results should be consistent with problem formulation. For example, for a linear static analysis the displacements should be 'small', and the material behaviour should remain within elastic limits.
- For problems with symmetry, the results should be symmetric.
- Displacements: check not only the form but also the magnitude of displacements.

Commercial codes, at least the good ones, also make checks on the condition and suitability of the stiffness matrix prior to solving the system equations. These checks are looking for two very common problems: *singularities* and *ill - conditioning*.

Ill-conditioning occurs when there is a large difference in the magnitude of the terms of the stiffness matrix. In this case, the solution would be prone to numerical round-off errors. The most common cause of ill conditioning is a large difference in the stiffness of parts of the model. This may be as a result of substantial differences in element size and/or material properties, or even large stiffness differences in a single element. To confirm that ill-conditioning is present, a reaction force check should be conducted.

Singularities occur when an non-unique or indeterminate solution is possible. Given a singularity, the system may do one of two things. In the first case, the analysis may abort. In the second, the analysis may continue with messages advising of negative or zero main diagonal terms.

The most common causes of singularities are (1):

- An unconstrained structure, where the complete model or any part connected through non-linear elements is not adequately constrained.
- An unconstrained joint. For example, two collinear horizontal spar elements will have an unconstrained vertical degree of freedom at the joint.
- An incomplete specification of real constants. For example, omission of some of the thicknesses at the nodes of shell or plane stress elements will lead to a singularity.
- Incorrect material properties. Obvious examples of this are zero elastic modulus or thermal conductivity.

Because of finite word size on computers, some singularities may appear as ill conditioning of the matrix, allowing the solution of the equations. However, such ill-condition results in large rigid body motions with relatively accurate reaction force checks.

4.4 Analysis Specification

This section considers some of the aspects of 'good practice' as applied to any Finite Element Analysis. It is good practice to draw up a formal specification for all but trivial analyses. These trivial investigations are commonly used as precursors to more elaborate exercises, and their function is usually confined to a specific aspect of the modelling problem. In any case, the format and detail required will depend on the purpose of the analysis, the scale of the analysis, and its use as the basis of an agreement between the analyst and customer.

At the very minimum, the specification should specify the objectives of the analysis, the level of checking and validation required, and the deliverables. It may include sources of data, milestones at which the analysis is reviewed, target or mandatory dates, and cost bounds.

There are some items that, if they can be defined in advance, should be specified. For example, the type of analysis, the software to be used, the extent of the model, the boundary conditions to be assumed, the load cases and combinations, the accuracy required, and the provision for re-analysis following design changes. It is also a good idea to include any numerical data that is available for such things as, material properties, loads and displacements and damping,.

Some of these items may depend on the outcome of previous stages of analysis and, therefore, their inclusion is left to the discretion of the analyst. A good example of specification is given in (5).

5. KEY ASSUMPTIONS IN FEA FOR DESIGN

Assumptions are an important and necessary part of simulation or modelling in order to quantify the physical problem under consideration. It is essential that the user understands the implications of the assumptions made. There are four major assumptions, which must be considered in any finite element based solution (6):

5.1 Geometry

Geometry should be considered in its proper context. Remember that the Computer Aided Design (CAD) geometry is only a template for building a mesh. Numerical analysis solvers have no capability to understand geometry as it is thought of in CAD. An FEA solver understands only nodes, and the connectivity of nodes, which are the elements. The smaller the element size or the higher the element order, the better the mesh will represent the geometry template. The inherent assumption when a model is solved is that the mesh represents the geometry adequately for the goals of the analysis. It is common practice to idealize the geometry by suppressing or removing unnecessary features or by representing solid structures as surfaces or lines. It is the user's responsibility to capture the geometrical features that are necessary for a correct analysis.

5.2 Material Properties

Specifying a single set of material properties makes the major assumption that all physical regions in the analysis have the same properties. It is also usual to assume that most regions will be isotropic and homogeneous. Such assumptions can limit the scope of the analysis. Recall that in Chapter 2, most solders were described as two phase alloys, with microstructures determined by local cooling rates, and were based on tin which is an anisotropic material!

Many responses to changing properties cannot be quantified very easily. It is advisable to compute the solution at several combinations of properties or bracket the solution with extremes of property values.

5.3 Mesh

The mesh is the means of communicating geometry to the FEA solver. The accuracy (not correctness) of the solutions is primarily dependent on the quality of the mesh, which is best characterized by the convergence of the problem. The global displacements should converge to a stable value, and any other results of interest should converge locally.

From a qualitative or visual point of view, a good-looking mesh should have well-shaped elements. Equilateral triangles and squares are the ideal. Although a good-looking mesh is not necessarily the best mesh, a bad-looking mesh almost always indicates a problem.

Transitions between mesh densities should be smooth and gradual without thin, distorted elements. The decision that a mesh is ready to solve, involves assuming that it will accurately represent the stiffness, or other properties of interest, of the intended structure.

5.4 Boundary Conditions

Choosing boundary conditions requires the greatest leap of faith for new users. Using boundary conditions to represent parts and effects that are not, or cannot be, explicitly modelled makes a tremendous assumption that the effects of these un-modelled entities can truly be simulated. Again the onus is on the user to interpret the results in light of the assumptions made. *The safest option is to assume that all results are wrong, unless proven otherwise.*

6. FUNDAMENTALS OF THERMAL ANALYSIS SIMULATION IN MICROELECTRONICS

6.1 Introduction

As heat, both its generation and removal, is a dominant condition in microelectronics, its investigation from a FEA point of view is worthy of attention. In this section we concentrate on conduction, although convection and radiation can play a role, depending upon the application.

Some of the basic principles and their FEA implementation are now reviewed. Heat transfer is the energy transfer process resulting from a temperature difference. A thermal analysis is undertaken to predict temperatures and heat transfer within and around bodies. This information may then be used to model temperature dependent phenomena, e.g. thermally induced strains and stresses, which in turn profoundly affect performance reliability.

There are three basic modes of heat transfer:

- Conduction - represents the internal energy exchange between one body in perfect contact with another, or from one part of a body to another due to a temperature gradient.
- Convection – represents the energy exchange between a body and a surrounding fluid.
- Radiation – represents the energy transfer from a body or between bodies by electromagnetic waves.

6.2 The Nature of Thermal Analysis

Whenever the specific heat (capacitance) matrix, thermal conductivity matrix, and/or equivalent nodal heat flow vector are functions of temperature, the analysis is non-linear and the equilibrium equations must be solved iteratively. Any of the following causes the analysis to be non-linear:

- Temperature-dependent material properties
- Temperature-dependent film coefficients
- Use of radiation elements
- Temperature-dependent heat sources (flow of flux)
- Use of coupled-field elements (assuming load vector coupling)

When the flow of heat does not vary with time, heat transfer is referred to as *steady state*. If the flow of heat does not vary with time, the temperature of the system and the thermal loads on the system also do not vary with time. From the First Law of Thermodynamics, the steady-state heat balance can be simply expressed as:

Energy in – Energy out = 0

Considering material properties alone, their sensitivity to temperature indicates that thermal analysis of the interconnection problem will be non-linear.

6.3 Thermal Loads and Boundary Conditions

As in all modelling, the load and boundary conditions are extremely important. The most common thermal loads and boundary conditions with respect to a finite element analysis are now described briefly. The term *load,* as used here, is generic for an applied influence, not just a mechanical force.

Temperature is a degree of freedom (DOF) constraint on a specific region of a model used to impose a known, fixed temperature. Although frequently used, this value is rarely known. The implication for using a fixed temperature as a boundary condition is that the system will supply or remove heat, as necessary, to maintain the temperature constant.

Uniform temperature can be applied to all nodes that do not already have a temperature constraint. This can only be used to set an initial temperature, not a constraint, on all nodes in the first substep of a steady-state or transient analysis.

The heat flow rate is a concentrated nodal load. A positive heat flow rate indicates energy is being supplied to the model, while a negative value indicates that heat is being extracted from the model. This type of load is typically applied when convection and heat flux cannot be used.

Convection is a surface load applied on the exterior surface of a model to simulate heat transfer between a surface and surrounding fluid.

Heat flux is also a surface load. It is used when the heat flow rate across a surface is known.

The heat generation rate is applied as a body load to represent heat generated within a body, and has units of heat flow rate per unit volume.

6.4 General Considerations

To summarise, thermal loads may be grouped into four general categories:
- DOF Constraint - specified DOF (Temperature) value
- Force Load – concentrated load (heat flow) applied at a point
- Surface Load- distributed load (convection, heat flux) over a surface
- Body load- volumetric or field load (heat generation)

Boundaries that have no applied loads are usually treated as adiabatic (perfectly insulated). Letting boundaries be adiabatic imposes symmetry boundary conditions. If a temperature of a region of the model is known, then it can be fixed at that value.

7. CLASSIFICATION OF HEAT CONDUCTION PROBLEMS

Heat conduction problems may be classified into four distinct groups. Depending upon the nature of a given problem, it is not uncommon to perform two or more of the analysis types. In general, it is a good idea to perform a basic linear steady state analysis first to get a "feel" for the problem and then proceed with the more realistic analysis. The four classes of analysis are (7):
- Linear steady state analysis
- Linear transient analysis
- Non-linear steady state analysis
- Non-linear transient analysis

7.1 Linear Steady State Analysis

In the steady state condition, a solid body is in thermal equilibrium with its surroundings. The one-dimensional governing differential equation with spatially dependent thermal conductivity $k(x)$ and spatially varying internal heat generation $Q(x)$ is given by:

$$-\frac{d}{dx}\left[k(x)\frac{dT}{dx}\right] = Q(x) \tag{16.14}$$

After finite element discretization, the resulting system of equations can be represented in matrix form as:

$$\left[K(x)\right]\{T\} = \{F(x)\} \tag{16.15}$$

With the conductivity matrix $K(x)$ and the flux vector $F(x)$ functions of spatial coordinates only. Within a single material, the conductivity, internal heat generation, convective heat transfer coefficient and heat flux value are either constant or a polynomial function of space only, i.e. not a function of time or temperature.

7.2 Linear Transient Analysis

A transient analysis implies that there is time dependence. If the solid body is subjected to some form of environmental change (e.g. allowed to cool) then the temperature of the body reduces until it reaches equilibrium with its surroundings. Thus, at every time instance, there is a heat exchange between the body and its surroundings, until the system reaches thermal equilibrium. Therefore, transient behaviour has a temperature distribution for every time instance.

Among the infinite number of time instances, the temperature distribution is calculated only at a discrete number of time steps. A transient or unsteady state analysis means that the body has NOT reached thermal equilibrium with its surroundings. The governing equation (16.14) is modified to take into account this change in internal energy with time as follows:

$$\rho C \frac{\partial T}{\partial t} - \frac{\partial}{\partial x}\left[k \frac{\partial T}{\partial x} \right] = Q(x,t) \qquad (16.16)$$

In a transient analysis, it is important that initial temperatures and spatial temperature prescriptions are mutually compatible. Any large differences that exist will need to be reflected in mesh refinement. For example, specifying two vastly different temperatures on two adjacent nodes would result in a very steep gradient within the element. The solution is to go to a higher order polynomial or, better, to refine the mesh in this area.

After discretization of (16.16) the matrix form is given by:

$$\left[C(x,t) \right]\left\{ \frac{dT(t)}{dt} \right\} + \left[K(x) \right]\left\{ T(t) \right\} = \left\{ F(x,t) \right\} \qquad (16.17)$$

In comparison with steady state, a transient analysis additionally requires the specification of density and specific heat capacity. An initial condition to satisfy the first term on the left-hand side of the governing equation (16.16) and specification of spatial conditions for the second term are also necessary.

In a transient analysis the temperature and flux vector are time dependent. There is also an additional time derivative term and a capacitance matrix. This matrix contains the product of density and specific heat capacity.

7.3 Non-linear Steady State Analysis

For heat transfer problems, non-linearity is characterized by temperature dependence, either in terms of material properties or boundary conditions. The one-dimensional governing equation for linear steady state problems (16.14) is modified as follows to incorporate this non-linearity.

$$-\frac{d}{dx}\left[k(x,T)\frac{dT}{dx} \right] = Q(x,T) \qquad (16.18)$$

The matrix form of the resulting systems of equations can be represented as:

$$\left[K(x,T)\right]\{T\} = \left\{F(x,T)\right\}$$ (16.19)

Thus, in a non-linear steady state analysis, the conductivity matrix and flux vector are functions of both spatial coordinates and temperature.

7.4 Non-linear Transient Analysis

A non-linear transient analysis is characterised by spatial, temperature and time dependence. Comparing the linear transient situation (16.17) to the non-linear case, equation 16.20 indicates that the capacitance matrix, conductivity matrix and flux vector are additionally dependent on temperature.

$$\left[C(x,t,T)\right]\left\{\frac{dT(t)}{dt}\right\} + \left[K(x,T)\right]\{T(t)\} = \left\{F(x,t,T)\right\}$$ (16.20)

Another characteristic of non-linear transient analysis is the non-linear boundary condition.   For example, radiation or convective heat transfer, where the heat transfer coefficient is a function of temperature and/or time.

An important application area for non-linear transient analysis is material phase change.   For example, solder bump formation.   During phase change, the heat capacity may change with respect to temperature.   For a more detailed coverage of the issues involved in FEA of heat transfer problems the interested reader is referred to (8).

8. SUMMARY

FEA is essentially an approximate method for calculating the behaviour of a real structure by performing an algebraic solution on a set of equations describing an idealized model structure with a finite number of variables. In this model the real structure is represented by a set of elements bounded by a mesh or grid of lines and surfaces. Each element is defined by its boundary geometry, material properties and a few basic parameters, e.g. thickness and cross-section area.

An element's behaviour in relation to adjoining elements is fully described by its boundary loads and displacements, which in turn are assumed to be functions of a finite number of discrete variables, nominally defined at the nodes or boundaries. The behaviour of the complete, idealized structure is determined as the aggregate behaviour of it elements.

The basic equations of equilibrium, compatibility and state are set up and solved in terms of the discrete boundary variables and the behaviour within elements is then derived from the values calculated at their boundaries.

Thus any FEA is only as good as:

- The model of the structure, i.e. the mesh
- The assumptions embedded in the properties used for each element
- The representation of the external loads and constraints in terms of the discrete boundary variables

Thermal issues are extremely important in microelectronics simulation. The user must understand the specific requirements of the various forms of thermal analysis. For example, to determine the form of analysis there are two key questions. Are the material properties or boundary conditions temperature dependent, and is there time dependence? There are four possible outcomes depending upon the two responses given. Thus, a No/No response implies a Linear Steady State Analysis; No/Yes a Linear Transient Analysis; Yes/No a Non-linear Stead State Analysis; and finally a Yes/Yes a Non-linear Transient Analysis. All four of these forms of analysis are applicable to solder simulation. The choice is a function of the problem to be solved. However, as stated earlier in this chapter, it is usually good practice to perform a linear static analysis first, no matter what the final analysis will be. This is often very quick, i.e. no non-linear material or transient behaviour to account for, and can point out several aspects of the model that may be incorrect or simply in need of special attention in the final model.

9 REFERENCES

1   MJ Fagan, *Finite Element Analysis: Theory and practice*, (Longman Scientific & Technical, UK), 1992

2   RD Cook, DS Mackus, ME Plesha, RJ. Witt, *Concepts and applications of Finite Element Analysis*, (Wiley), 2002

3   OC. Zienkiewics, RL Taylor, *The Finite Element Method*, Fifth Ed., 3 Volumes, (Butterworth Heinemann), 2000

4   D Baguley, DR Hose, *How to – Model with Finite Element,* NAFEMS, 1997

5   D Baguley, DR Hose, *How to – Plan a Finite Element,* NAFEMS, 1994

6   V Adams, A Askenazi, *Building Better Products with Finite Element Analysis*, (Onword Press, 1999)

7   RS Ransing, SJ Hardy, DT Getting, *How to – Undertake Finite Element Based Thermal Analysis*,NAFEMS, 1999

8   RW Lewis, K Morgan, HR Thomas, KN. Seetharamu, *The Finite Element Method in Heat Transfer Analysis*, (Wiley, UK), 1996

CHAPTER 17

NON-LINEAR FINITE ELEMENT ANALYSIS

1 INTRODUCTION

Recall from Chapter 16 that the basic principle of the finite element method is that a mathematical model of a structure can be built up from a series of discrete 'finite' elements, each of which has an analytically defined relationship between force and displacement (in a mechanical problem). This force-deformation relationship, or stiffness, of each element can be assembled to create a global stiffness matrix for the whole structure. Then, by employing a matrix solution technique, the response of the entire structure to the prescribed loads and boundary conditions can be determined.

The finite element method is characterised by the simple relationship between a vector of forces, F, the stiffness matrix, K and a displacement vector, U where

$$\{F\} = [K]\{U\} \tag{17.1}$$

The majority of the FE analysis performed to date has been linear; that is, the stiffness, K, of the structure being analysed is assumed to remain constant. In most cases, this is an approximation, since even small deflections in a structure are likely to introduce some changes in stiffness. However, in many instances, the approximation is valid within the limits of the engineering analysis and so the results are quite acceptable.

This assumption of linearity, in the FE solution, provides several key benefits. For example, the results can be factored – if a load of 10N produces a deflection of 5mm in a structure, a load of 25N will produce a deflection of 12.5mm. In this case the past loading history of the structure is not important – 'residual' stresses can be ignored. Another major benefit is the principle of superposition – results obtained from the individual application of different loads can be factored and added together to produce the results from the simultaneous application of all the loads. The factoring and superposition of results is particularly useful when analysing an object where many separate loads may be applied, in different combinations and with different magnitudes.

Despite the clear benefits of linear FE analysis, the assumption of linearity can severely limit the capacity of the finite element technique to model real engineering problems undergoing realistic loading. A major drawback of the assumption that the structural stiffness remains constant during the analysis is the implication that the material properties must remain constant throughout the analysis. This precludes the analysis of any structure in which the material might yield and undergo plastic deformation or in which the material stiffness is dependent on the strain, the strain rate or a temperature field, which vary with time. All these conditions are likely to exist during the life of a soldered interconnection.

305

There are other limitations inherent in the linear assumption. For example, changes in the structural stiffness that may arise from contact between parts of the structure and/or any subsequent frictional sliding cannot be taken into account in a linear analysis. Also, linear analysis will become invalid if the structure undergoes large deflections, because significant changes in the shape or orientation of a structure can alter its stiffness.

In general, it can be seen that many real-life engineering problems, including micro-electronics with its different materials, high working temperatures over long periods and other conditions varying with time, cannot be analysed accurately using a linear FE approach because there are too many non-linearities for the assumption to remain valid.

Given these limitations, how can a non-linear analysis overcome such problems? In simple terms, a non-linear finite element analysis differs from a linear analysis in that the 'stiffness' of the structure in a non-linear analysis is not assumed to be constant throughout the simulation. That is to say, $F \neq K U$, and we consider the stiffness to include both geometric, load and material stiffness terms. It will be shown in subsequent sections, that this change in the underlying assumptions of the finite element method has an effect on the way the input data are defined, the way the solution is obtained, and the way in which the results are interpreted.

In some cases, the non-linear model may be more difficult to set up; it may be more computationally intensive to run, and it may require the interpretation of a larger quantity and a wider range of results

Fundamentally, non-linear analysis is an extension of linear analysis: it is based on matrix solution techniques and employs elements, material models, loads, and boundary conditions. There is an important difference, however, and that is the solution procedure is an incremental-iterative process and the elements and constitutive laws may be highly sophisticated.

It is important to note that both linear and non-linear FE analyses should provide the same answer for a linear problem. It is only when the problem contains some form of non-linearity that the linear assumptions lead to a loss of accuracy.

Table 17.1 illustrates the main differences between linear and non-linear analyses.

Table 17-1. Comparison between linear and non-linear analysis (1)

Feature	Linear	Non-linear
Load-displacement	Displacements are linearly dependent on the applied loads.	The load-displacement relationships are non-linear.
Stress-strain relationship	A linear relationship is assumed between stress and strain.	With material non-linearity, this relationship is a non-linear function of stress, strain and/or time.
Magnitude of displacement	Changes are assumed small and ignored. The original (undeformed) state is used as the reference state.	Displacements may not be small; hence an updated reference state may be needed.
Material properties	Linear elastic properties are usually easy to obtain.	Non-linear properties may be difficult to obtain.
Reversibility	The behaviour of the structure is completely reversible upon removal of the external loads.	Upon removal of external loads, the final state may be different from the initial state.
Boundary conditions	Remain unchanged throughout the analysis.	May change, e.g. a change in contact area.
Loading sequence	Not important. The final state is unaffected by the load history.	The behaviour of the structure may depend on the load history.
Iterations and increments	The load is applied in one load step with no iterations.	The load is often divided into small increments with iterations performed to ensure that equilibrium is satisfied at every load increment.
Computation time	Relatively small in comparison to non-linear problems.	Due to the many solution steps required for load incrementation and iterations, computation time is high.
Use of results	Superposition and scaling allow results to be factored and combined as required.	Factoring and combining of results is not possible due to the history dependence.
Initial state of stress/strain	Unimportant.	Usually required for material non-linearity problems.

2 TYPES OF NON-LINEARITIES

There are three major types of non-linearities typically used in engineering problems. The most relevant to microelectronics is material non-linearity, which can be can further divided into two distinct types of behaviour:

- Time-independent behaviour, such as the elasto-plastic behaviour of solder loaded past the yield point.
- Time-dependent behaviour, such as the creep of solder when the effect of change of strain with time is important. This usually requires some form of power law to express the stress-strain rate relationship. This was covered in Chapter 3. At fixed strain, stress varies with time as stress relaxation occurs. Two other material models of this type are viscoelastic and viscoplastic behaviour in which both effects of plasticity and creep are combined. In this case, the stress is dependent on the strain rate.

The other two non-linearities include: Geometric Non-Linearity, which covers large deformations and/or large strains; and Boundary Non-Linearity, for example the opening/closing of gaps and contact. These will be discussed in more detail in subsequent sections.

3   MATERIAL NON-LINEARITIES

The assumptions of linearity are based on the practical experience that for small deformations, many materials exhibit an elastic response – the material returns instantaneously to its original shape when the loading is removed. In addition the value of the elastic stiffness, Young's modulus, is a constant for a particular material at a particular temperature. (See Chapter 2 for more detail).

The use of the term 'small' raises the obvious question: 'How small is small?' There is no definitive answer. In fact, it is better to use the term "significant", since the decision on whether a number is small or not depends largely on whether its magnitude is significant in terms of the analysis being performed.

In analysing the behaviour of a structure when the strains in the material are not small and the material no longer behaves elastically, the original shape is not retained after the loading is removed. This behaviour is normally described as plasticity, and is characterised by a stress-strain curve where the material has a constant elastic stiffness up to a certain small strain value and then undergoes some form of yielding, followed by plastic flow at higher values of strain. Further details were presented in Chapter 2.

Such material behaviour cannot be modelled using the linear finite element method because once the yield point is reached the stiffness changes as the deformation increases. A non-linear solution technique must, therefore, be used.

Many metallic materials have clearly defined yield points, so that data required to characterise behaviour can be as simple as Young's modulus, Poisson's ratio, yield stress, and hardening modulus. However, most ductile materials may exhibit strain rate dependence, where the values of the yield stress and hardening modulus change depending on the rate at which the materials are being deformed. In addition, as seen earlier, yield stress and hardening modulus may also be highly dependent on temperature. With their generally small yield stress, and sensitivity to strain rate and temperature, solders are most unlikely to be amenable to linear analysis.

3.1 Time-Independent Behaviour (Plasticity)

When a ductile material experiences stresses beyond the elastic limit, it will yield, acquiring large permanent deformations. Plasticity refers to the material response after yield. In metals, such plasticity was described in Chapter 2. The following sections consider the process from a more analytical viewpoint, with an emphasis on the finite element method.

3.1.1 Outline of a Finite Element plasticity algorithm

There are several assumption used in modelling elasto-plasticity. These may be summarised as follows:

- Elasto-plastic flow is path-dependent, i.e. it is dependent on the load path, but is independent of time.
- All of the variables, e.g. displacement, can be expressed as 'rates', where rate is used to indicate a sequence of events, as opposed to change with time. Time in this form of analysis, is referred to as pseudo-time.
- The external loads are applied in small increments and the deformation and resulting stresses can be accumulated to give the final state of stresses and strains.

In elasto-plastic problems, the use of load incrementation and equilibrium iterations is necessary to achieve the final solution. Within each load increment, the plasticity relationships, which depend on the current state of loading, are satisfied and the variables updated for the next load increment. Iterations are required within each increment in order to ensure that the requirements of satisfying the constitutive equations and the overall equilibrium are satisfied. The iterations are terminated when the solution has converged to some previously specified criteria. The number of iterations depends on the size of the load increment, and whether the stiffness matrix has been updated to reflect the current stress-strain state.

A simplified FE algorithm for modelling elasto-plastic behaviour is given by the following (1)

- Apply the load assuming elastic behaviour everywhere. The overall stiffness matrix is obtain, and the following equations solved to obtain the displacement vector:

$$[K_e][u] = [F] \qquad (17.2)$$

where $[K_e]$ is the elastic stiffness matrix, and $[F]$ is the full load vector.

- Calculate the effective stress at all Gauss (integration) points, and check the maximum value against yield stress, σ_y. If it does not exceed σ_y, then there is no need for a plasticity analysis. If it does, scale down the magnitudes of all nodal displacements such that the node or Gauss point with the highest effect stress is just yielding. The scaling factor is used to determine the fraction of the applied loads that causes initial yielding.

- Divide the remainder of the applied load into small increments, either as specified by the user, or according to a suitable automatic scheme.
- Apply one load increment, and re-solve the equations to obtain the new displacements corresponding to this load vector. Use either the initial (elastic) stiffness matrix $\left[K_e \right]$, or the tangent stiffness matrix, $\left[K_p \right]$, which is updated to contain the current state of plasticity, as follows:

$$\left[K_p \right]\left[\Delta u \right] = \left[\Delta F \right] \tag{17.3}$$

Many Finite Element codes use the initial stiffness matrix and then update it after a few load increments or iterations to reflect the current state of plasticity.

- Perform iteration to ensure that the solution is acceptable, i.e. it satisfies both the equilibrium conditions and the plasticity material laws. From the computed $\left[\Delta u \right]$, the total and plastic strain increments can be calculated, and then the corresponding stress. An out-of-balance or residual force vector, $\left[R \right]$, is calculated by integrating the stresses over all the elements and subtracting the internal forces from the external forces. If convergence is not achieved, i.e. $\left[R \right]$ is not smaller than a specified tolerance, an iteration is commenced by solving a new set of equations, as follows:

$$\left[K_p \right]\left[\Delta u_{corr} \right] = \left[R \right] \tag{17.4}$$

Where $\left[\Delta u_{corr} \right]$ is a correction to the displacement vector in order to balance the residual force vector. Using this correction, the displacement vector is improved and the iteration repeated to obtain a new residual force vector. The displacement corrections and the residual forces should be getting smaller with each iteration, until convergence is achieved.

- Store the computed increments of displacements, strains and stresses at each node, and update the existing values.
- Apply the next load increment and perform iterations as necessary. Terminate the calculations when the final load is processed.

Summarising the important points in the above algorithm, the fundamental characteristic of a non-linear FE solution is that the stiffness matrix is not constant during the analysis. The stiffness matrix may need to be recalculated many times over the course of the analysis; therefore, an iterative-incremental solution technique is normally employed. The loading is applied in successive incremental stages; and an iterative scheme is used to find equilibrium at each increment. In this way the solution proceeds with successive changes to the stiffness at each increment until the final load has been reached.

Several consequences arise from the use of this solution procedure: The non-linear analysis almost certainly takes longer than a linear analysis of the same model because the matrix solution is carried out many times rather than just once. In addition, there is the possibility that the analysis will not complete at all. The incremental, iterative approach relies on the non-linear changes being relatively smooth. If there are sharp changes in the stiffness of the structure, the solution may not converge to an equilibrium state.

The level of expertise of the analyst, and the quality of the software itself are often key factors that determine whether or not a non-linear analysis is successful. A very good and comprehensive description of Finite Element algorithms and procedures used in non-linear analysis is presented by (2).

3.2 Time Dependent Behaviour (Creep, Viscoelasticity, Viscoplasticity)

In this class of material models, rate refers to real time, unlike the models described above.

3.2.1 Creep

Creep is a rate (time) dependent material non-linearity in which the material continues to deform under a constant load. If a uniaxial tensile specimen is loaded at high temperature (T>0.4 Tm, where units are in Kelvin) the change of strain as a function of time, at a constant temperature, is similar to the form given in Figure 17-1.a. As presented in Chapter 2, the time-strain curve may be divided into three stages termed primary, secondary and tertiary creep, although some authors define four stages to include the initial elastic strain on loading.

In the primary stage, the strain rate decreases with time. The secondary stage has a constant or approximately constant, strain rate associated with it; see Figure 17-1.b. While in the tertiary stage, the strain rate increases rapidly until rupture. Creep behaviour can be represented by a set of curves relating the stress, strain and time. Those for constant time (isochronous) and constant strain (isostrain) are illustrated in Figure 17-1.c.

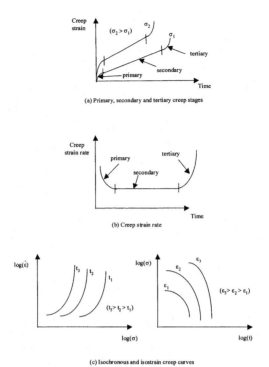

Figure 17-1 a) Primary, Secondary and tertiary stages of creep, b) creep strain rate, c) Isochronous and isostrain creep curves.(1)

The dependency of creep deformation rate $\dot{\varepsilon}_c$ as a function of stress, strain, temperature and time, can be represented in a form similar to:

$$\dot{\varepsilon}_c = f_1(\sigma)f_2(\varepsilon)f_3(T)f_4(t) \tag{17.5}$$

Although this form of equation has proved to be useful, it is the user's responsibility to insure, via experimental evidence, that it is applicable to a particular material and stress-temperature range. More detailed information on these and other models is available elsewhere (3).

The stress dependence of secondary (or minimum) creep has been described in several ways. Two of the most common are:

$$\dot{\varepsilon}_c = A\sigma^n \quad \text{Norton} \quad , \dot{\varepsilon}_c = B\sinh(\beta\sigma) \quad \text{Prandlt}$$

The time dependence of creep, equation 17.5, has been expressed in several forms, including:

$$f_4(t) = t \quad \text{Secondary} , \quad f_4(t) = bt^m \quad \text{Bailey}$$

The exponent m<1 is used to represent the primary creep stage.

The temperature dependence, from both theoretical considerations and experimental evidence, is of the form:

$$f_3(T) = \exp\left(\frac{-Q}{RT}\right) \tag{17.6}$$

where Q is the activation energy, R the Boltzmann constant, and T is the absolute temperature. Refer to Chapter 3.

A favourite model to represent secondary creep in isothermal conditions is the Norton model. This is given by:

$$\dot{\varepsilon}_c = A\sigma^n t \tag{17.7}$$

where A and n are material constants determined experimentally. If both primary and secondary behaviour are of interest, then the Norton-Bailey model is often used,

i.e. $$\dot{\varepsilon}_c = A\sigma^n t^m \tag{17.8}$$

Alternatively, the secondary creep and Bailey models may be combined,

$$\dot{\varepsilon}_c = A_1 \sigma^{n1} t + A_2 \sigma^{n2} t^m \tag{17.9}$$

Again, it is the user's responsibility to insure that the model chosen from the many available is applicable to the problem at hand. However, most Finite Element codes consider only the primary and secondary stages of creep.

A change in stress during primary creep can be accommodated in modelling by assuming either a *time hardening* or *strain hardening* response. In the time hardening approximation, the creep strain rate depends only upon the time from the beginning of the creep process. That is to say, the stress curve shifts up or down. As the stress changes from σ_1 to σ_2, the different creep rates are calculated at points A to B.

Strain hardening assumes that the creep strain rate depends only on the current accumulated creep strain. In this case, the stress curve shifts left or right. As the stress changes from σ_1 to σ_2, the different creep rates are calculated at points A to B.

The differences between the time hardening and strain hardening assumptions can be illustrated as in Figure 17-2 (4). The time hardening assumption is easier to use, but the strain hardening assumption is considered more accurate. Neither time nor strain hardening assumption is suitable for situations where the stress changes sign, that is stress reversal or in covering all three stages of creep. Both of these assumptions are usually implemented in commercial Finite Element codes.

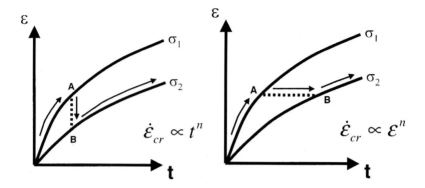

Figures 17-2 (a) time hardening (b) strain hardening

In Table 17.2 the various forms of implicit creep models, as supported by the commercial, ANSYS® Finite Element, code are illustrated. Although this list is not exhaustive, it is representative. Note that ANSYS® does not directly support tertiary creep models.

Table 17-2 An example of ANSYS ®implicit creep models

Stage	Name	Equation
Primary creep	strain hardening	$\dot{\varepsilon}_c = C_1 \sigma^{C_2} \varepsilon^{C_3} e^{-C_4/T}$
Primary creep	time hardening	$\dot{\varepsilon}_c = C_1 \sigma^{C_2} t^{C_3} e^{-C_4/T}$
Primary creep	generalized exponential	
Primary creep	generalized Graham	$\dot{\varepsilon}_c = C_1 \sigma^{C_2} \left(t^{C_3} + C_4 \sigma^{C_5} + C_6 \sigma^{C_7} \right) e^{-C_8/T}$
Secondary creep	generalized Garofalo	$\dot{\varepsilon}_c = C_1 \left[\sinh \left(C_2 \sigma \right) \right]^{C_3} e^{-C_4/T}$
Secondary creep	exponential form	$\dot{\varepsilon}_c = C_1 e^{\sigma/C_2} e^{-C_3/T}$
Secondary creep	Norton	$\dot{\varepsilon}_c = C_1 \sigma^{C_2} e^{-C_3/T}$
Primary + secondary creep	time hardening	$\dot{\varepsilon}_c = \dfrac{C_1 \sigma^{C_2} t^{C_3+1} e^{-C_4/T}}{\left(C_3 + 1 \right)} + C_5 \sigma^{C_6} t e^{-C_7/T}$

3.2.2 Viscoelasticity

Viscoelasticity is a material non-linearity that has both an elastic (recoverable) part of the deformation, as well as a viscous (non-recoverable) part. Upon application of the load, the elastic deformation is instantaneous while the viscous part occurs over time. It can occur in all types of materials under the appropriate conditions, although, as indicated in Chapter 2, it is most common in polymers above their glass transition and in metals at high homologous temperatures.

Viscoelasticity accounts for the effect of the time-rate of change of stress in the material, so that a viscous component to the elastic straining is included. Temperature effects are also included in this model, so heating and cooling sequences may be simulated. The viscoelastic model is used to characterize the behaviours of glass or glass-like materials.

Since viscoelasticity is a combination of elastic solid and viscous liquid behaviour the user must refer to the code used to ascertain the appropriate input data required to capture this behaviour. However, it will certainly involve a substantial number of parameters that must be determined experimentally. For a recent example of viscoelastic modelling the reader is referred to (5).

3.2.3 Viscoplasticity

Viscoplasticity is a time-dependent plasticity phenomenon, where the extent of the development of the plastic strains is dependent on the rate of loading. It is primarily applicable in situations involving large plastic and small elastic strains, such as metal forming (hot working) and soldered interconnections. The viscoplastic model for solder (6, 7) unifies plasticity and creep via a set of flow and evolutionary equations, and has been the basis of much modelling in microelectronics (8, 9,10). The following form for the flow equation of the Anand model is given as follows:

$$\dot{\varepsilon}_{in} = Ae^{-\frac{Q}{RT}}\left[\sinh\left(\xi\frac{\sigma}{s}\right)\right]^{\frac{1}{m}}$$

(17.10)

Where $\dot{\varepsilon}_{in}$ is the inelastic strain rate, A the pre-exponential factor, Q the activation energy, m the strain rate sensitivity, ξ the multiplier of stress, R the gas constant, and T the absolute temperature, respectively.

The evolution equation for the internal variable, s , is derived as

$$\dot{s} = \left\{h_0\left|1-\frac{s}{s^*}\right|^a sign\left(1-\frac{s}{s^*}\right)\right\}\bullet\dot{\varepsilon}_p, a < 1$$

(17.11)

with

$$s^* = \hat{s} \left[\frac{\dot{\varepsilon}_p}{A} \exp\left(\frac{Q}{RT} \right) \right]^n \qquad (17.12)$$

where h_0 is the hardening/softening constant, a the strain rate sensitivity of hardening/softening, s^* the saturation value of s, $\hat{s}$ the coefficient, and n the strain rate sensitivity for the saturation value of deformed resistance.

There are nine material parameters in the viscoplastic Anand model, which can be determined directly from the experimental data, e.g. a series of creep and constant strain rate tests. Viscoplasticity is characterized by the irreversible straining that occurs over time as a function of the strain rate. From a material viewpoint, viscoplasticity and creep are equivalent. In Table 17.3, these nine parameters, their meaning and units are presented, while in Table 17.4 typical values for four different solders are given.

Table 17-3 Material parameters used for Anand model

Parameter	Meaning	Units
s_0	Initial value of deformation resistance	Stress, e.g. MPa
$\dfrac{Q}{R}$	Q = activation energy R = universal gas constant	Energy/volume e.g. kJ/mole Energy/(volume-temperature), e.g. kJ/(mole-°K)
A	Pre-exponential factor	1/time e.g. 1/second
ξ	Multiplier of stress	Dimensionless
m	Strain rate sensitivity of stress	Dimensionless
h_0	Hardening/softening constant	Stress, e.g. MPa
$\hat{s}$	Coefficient for deformation resistance saturation value	Stress, e.g. MPa
n	Strain rate sensitivity of saturation (deformation resistance) value	Dimensionless
a	Strain rate sensitivity of hardening or softening	Dimensionless, a >1.0

Table 17-4 Typical values for solders of parameters used in Anand model

Parameter	Pb-5Sn-2.5Ag	Sn-36Pb-2Ag	Sn-40Pb	Sn-3.5Ag
s_0	1.052(105)	2.30(107)	1.49(107)	2.23(104)
$\dfrac{Q}{R}$	11010	11262	10830	8900
A	7	11	11	6
ξ	0.241	0.303	0.303	0.182
m	41.63	80.79	80.42	73.81
h_0	0.002	0.0212	0.0231	0.018
$\hat{s}$	11432	4121.31	2640.75	3321.15
n	1.3	1.38	1.34	1.82
a	33.07	42.32	56.33	39.09

4 BOUNDARY AND GEOMETRIC NON-LINEARITIES

Although these two types of non-linearity may not be directly applicable to solder interconnection problems, they are presented here to complete the description of all forms of non-linearity.

Boundary non-linearity occurs in most contact problems, in which two surfaces come into or out of contact.   This type of non-linearity may occur even if the material behaviour is assumed linear and the displacements are infinitesimal, due to the fact that the size of the contact area is usually not linearly dependent on the applied loads. If the effect of friction is included in the analysis, then stick-slip behaviour may occur which adds a further non-linearity that is normally dependent on the loading history.

Geometric non-linearity is perhaps the most difficult to describe of the three categories. This non-linearity covers all situations where a change in the geometry or orientation of a structure leads to a change in its response. This does not necessarily arise from large displacements, since the structural stiffness could still be the same even though the component has moved a considerable distance and, in fact, non-linear geometric effects can be important when displacements are very small. Geometric non-linearities also arise when a structure undergoes large rotations, and when structural instabilities, such as snap-through and post-buckling behaviour occur.

Geometric non-linearities refer to the non-linearities in the structure or component due to the changing geometry as it deflects. That is, the stiffness [K] is a function of the displacements {U}. The stiffness changes because the shape changes and/or the material rotates.

5 SUMMARY

There are three major types of non-linearity: Geometric (large deformations, large strains), Material (plasticity, creep, viscoplasticity/viscoelasticty) and Boundary (contact). These may occur singly or in combination.

Indications of possible non-linear behaviour include permanent deformations and gross changes in geometry, cracks, necking, thinning, buckling, stress values which exceed the elastic limits of the materials, evidence of local yielding, shear bands, and homologous temperatures above 0.4. In these cases the stress is no longer proportional to the strain.

Non-linear problems are inherently more complex to analyse than linear problems. And the principle of superposition (which states that the resultant deflection, stress, strain in a system due to several forces is the algebraic sum of their effects when separately applied) no longer applies

In this module we reviewed the basic concepts of plasticity. That is when a ductile material experiences stresses beyond the elastic limit; it will yield and acquire large permanent deformations. An outline of the Finite Element plasticity algorithm was given. The important fact here is the necessary use of load incrementation and equilibrium iterations to achieve the final solution.

We next considered viscoplasticity, and in particular the Anand model. This is an important model for those wishing to unify plasticity and creep for solder models. The model requires a substantial number of parameters, i.e. 9, which must be determined from experimental work.

In conclusion, linearity is very useful, but for generally simple situations. Non-linearity is needed for a large proportion of real-life modelling – including solders.

6 REFERENCES

1   AA Becker, Understanding Non-Linear Finite Element Analysis Through Illustrative Benchmarks, NAFEMS, 2001.
2   E Hinton (Ed), NAFEMS Introduction to Nonlinear Finite Element Analysis, NAFEMS, 1992.
3   RK Penny, DL Marriott, Design for Creep, 2nd Ed., Chapman & Hall, 1995.
4   Basic Structural Non-Linearities, ANSYS, 2000 SAS IP, Inc.
5   MS Kiasat, et al, Time and Temperature Dependent Thermo-Mechanical Characterization and Modeling of a Packaging Molding Compound, in "Benefiting from Thermal and Mechanical Simulation in (Micro)-Electronics", EuroSimE2001, 2001.
6   L Anand, Constitutive equations for hot working of metals, Intl. J. Plasticity, 1, 213-231, 1985.
7   SB. Brown, HK Kim, L Anand, An Internal Variable Constitutive Model for Hot Working of Metals, Intl. J. Plasticity, 5, pp 95-130, 1989.
8   ZN Cheng, GZ Wang, L Chen, J Wilde, K Becker, Viscoplastic Anand model for solder alloys and its application, in Solder & Surface mount Technology, 12/2, 2000, 31-36.
9   R Darveaux, Solder Joint Fatigue Life Model, Proc. Of TMS Annual Meeting, 213-218, 1997.
10 R Darveaux, Effects of Simulation Methodology on Solder Joint Crack Growth Correlations, Proc. Of 50th Electronic Components & Technology Conference, 1048-1058, 2000.

CHAPTER 18

CASE STUDY

1 INTRODUCTION

Failure of solder interconnections under service conditions is a critical issue in electronic packaging and assembly. Thermo-mechanical modelling is seen as an efficient way to explore the behaviour of solder interconnections and improve solder reliability. There is a great deal of activity concerned with developing reliable finite element models for the prediction of thermal-mechanical behaviour of solder.

As an example of this, the present Chapter focuses upon the work of Zhao, *et al*, (1) presented at a recent conference on Simulation in Micro-Electronics.

This study concerns solder interconnections in CSP (Chip Scale Packages). There are three key factors that determine the accuracy of the solder interconnection reliability prediction.

1. The strong effect of the solder geometry on the thermo-mechanical behaviour of the solder interconnection
2. The viscoplastic constitutive characteristics of eutectic SnPb solder
3. A reliable 3D model

Given these factors, the thermo-mechanical behaviour of the solder interconnection can be predicted as a function of the solder geometry parameters. Clearly, the solder interconnection provides an important mechanical and electrical connection between the silicon chip and packaging component with the PCB.

Electronic products experience temperature fluctuations under service conditions and this may cause progressive damage in the solder joint. The accumulated damage may lead to failure. The ability to identify the failure mechanism of solder joints is extremely important. Commonly used FEA 2D and 3D models with fewer bumps than exist in practice may not work well as all solder bumps will contribute to the eventual interconnection failure. Accurate 3D geometric solder modelling is often neglected, which may lead to some doubt about the reliability of the modelling.

For widely used solder alloys, the kinetics of deformation are highly temperature and rate dependent. A constitutive model to describe the large viscoplastic deformations is required. Many constitutive equations have been proposed including: single creep equations with power-law or hyperbolic sine functions, elastic + plastic, elastic + plastic + creep. However, the inelastic behaviour of materials comes from the unified evolution of internal structure and it is not possible to separate it into plasticity and creep. Thus it is advisable to model solder allows using unified viscoplastic constitutive equations.

In these simulations, the formation of the solder joints is developed using the software "Surface Evolver". The predicted solder geometry is then transferred to ANSYS(R). A 3D finite element model is developed for a 5x5 CSP area array with the predicted solder geometry. The Anand viscoplastic model, as discussed in

Chapter 17, is used to describe the constitutive characteristics of the solder because it accounts for the unified nature of plasticity and creep.

2. SOLDER GEOMETRY MODELING

The formation of a solder joint generally involves a reflow process to heat a solder paste at the junction of the component and board until the solder liquifies. The solder is then cooled down to form the final joint and the component is thus interconnected with the pad. Based on minimum energy theory, the final solder geometry depends on a potential energy balance. If the contraction of solder volume in the process of solidification is ignored, the liquid solder will reach the equilibrium shape when the total potential energy in the molten solder-gas-component-PCB system is minimized. Figure 18-1 below shows a typical solder joint in CSP.

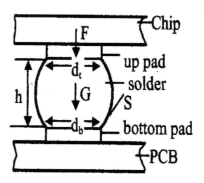

Figure 18-1 Typical solder joint in CSP

The mathematics of determining the surface shape may be found in the cited reference (1).

The ratio of the top and bottom pad diameters is a sensitive solder geometry parameter. The different solder geometries with variable d_b / d_t ratios are shown in Figure 18-2.

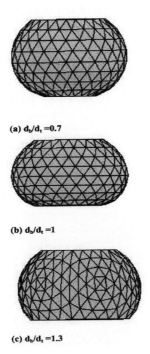

(a) $d_b/d_t = 0.7$

(b) $d_b/d_t = 1$

(c) $d_b/d_t = 1.3$

Figure 18-2 Predicted solder geometry with variable d_b / d_t ratios

Some parameters used in the simulation and the geometry characteristics are given in the table 18.1.

Table 18-1. Solder joint Geometry Parameters

d_b / d_t	0.7	1	1.3
Solder Volume (mm^3)	0.0185	0.0185	0.0185
d_t (mm)	0.22	0.22	0.22
Surface Tension (dyne/cm)	470	470	470
Density (g/cm^3)	831	831	831
Maximum Diameter (mm)	0.338	0.336	0.356
Height (mm)	0.26	0.236	0.228

3. MECHANICAL ANALYSIS

3.1 Geometry Model

The total geometry model consists of three components: Silicon chip, Solder joints, FR4 board. It has a 5x5 solder ball array at 0.5 mm pitch and is based on a 25mm square FR4 board. The chip size is 2.6 x 2.6mm square with a thickness of 0.65mm. Only one quarter of the packaging was modelled due to symmetry.

The solder joints were meshed with 8 node brick elements (VISCO107). The chip and PCB were meshed with 20 node solid elements (SOLID95). There are 10770 nodes and 11740 elements in the model. The 3D mesh for the CSP is given in Figure 18-3.

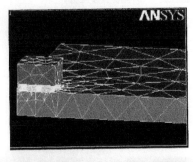

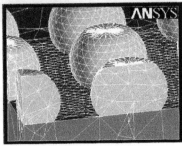

Figure 18-3 Mesh of the Finite Element Model for CSP

Loading consisted of 36°C/min temperature ramps followed by 10 min hold – time at the extreme temperatures. The temperature range was from –55°C to 125°C, with two cycles per hr. The loading history, as temperature profile, is given below in Figure 18-4.

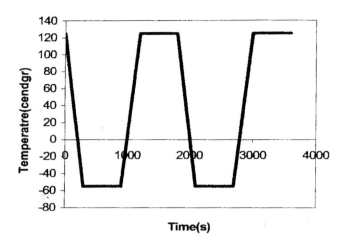

Figure 18-4 Temperature profile of thermal cycle imposed on solder joints

3.2 Material Model

The viscoplastic Anand model for large, isotropic, viscoplastic deformations, but small elastic deformations is a single-scalar internal variable model. Two important features of the Anand model are that: it needs no explicit yield condition and no loading/unloading criterion, and it employs a single scalar as an internal variable to represent the averaged isotropic resistance to plastic flow. This variable is denoted by "s", has units of stress and is called deformation resistance.

The deformation resistance, s is consequently proportional to the equivalent stress. That is,

$$s = \frac{\sigma}{c}, c < 1 \tag{18.1}$$

where, c is a material parameter obtained from a constant strain rate test.

The flow equation exactly accommodates the strain rate dependence on the stress,

$$\dot{\varepsilon}_p = A \exp\left(-\frac{Q}{RT}\right)\left[\sinh\left(\xi\frac{\sigma}{s}\right)\right]^{\frac{1}{m}} \tag{18.2}$$

where R is the gas constant, T is the absolute temperature and the other variables are listed in table 18.2.

The evolution equation (strain hardening/softening) for the internal variable s is derived as follows:

$$\dot{s} = \left\{ h_0 \left| 1 - \frac{s}{s^*} \right|^a sign\left(1 - \frac{s}{s^*} \right) \right\} \bullet \dot{\varepsilon}_p, a \langle 1 \tag{18.3}$$

With s^* the saturation value of deformation resistance, s, given by:

$$s^* = \hat{s} \left[\frac{\dot{\varepsilon}_p}{A} \exp\left(\frac{Q}{RT} \right) \right]^n \tag{18.4}$$

Where, in summary:

$\dot{\varepsilon}_p$	effective inelastic deformation rate
σ	effective Cauchy stress
s	deformation resistance
s^*	saturation value of deformation resistance
$\dot{s}$	time derivative of deformation resistance
T	absolute temperature

There are nine material parameters associated with the viscoplastic Anand model. In this particular case, Anand's constants, as proposed by (2) are used and are given in table 18.2. Related material parameters are given in table 18.3.

Table 18- 2 Anand's parameters for Sn-37Pb

Constants	Value	Units	Description
A	4.0e6	(1/time)	Pre-exponential factor
Q/R	9400	Energy/volume	Activation Energy/Universal Gas Constant
ξ	1.5	dimensionless	Multiplier of Stress
m	0.303	dimensionless	Strain rate sensitivity of stress
h_0	2.0e5	Stress (MPa)	Hardening/softening constant
$\hat{s}$	2000	Stress (MPa)	Coefficient for saturation value of deformation resistance
n	0.07	dimensionless	Strain rate sensitivity for the saturation value (deformation resistance)
a	1.3	Dimensionless a>1.0	The strain rate sensitivity of hardening/softening
s_0	1800	Stress (MPa)	Initial value of deformation resistance

Table 18-3 Material Properties

Material	Young's modulus (MPa)	CTE (ppm/cendgr)	Poisson's ratio
Solder	67180-108T	21	0.4
Silicon	131	2.8	0.3
FR4	24420-22.6T	17.6	0.28

3.3 Simplifications and Assumptions

Some simplifications and assumptions were applied to obtain a feasible finite element model. They were:

- No volume change occurs during solidification of the solder.
- No temperature gradient exists over the entire solder joint during reflow.
- There is no offset between the packaging component and substrate, that is, the solder joint is axisymmetric.
- Small strain anistropy associated with inelastic and Bauschinger type effects can be ignored.
- The solder joint is intact without any defects.
- Pads on components and PCB can be ignored; only interactions between component, solder and PCB are considered.

4 ANALYSIS RESULTS

Several thermal cycles are simulated and each cycle is divided into 4 load steps. An evenly distributed step scheme is adopted with 10 sub-steps / load step. According to the results, the maximum inelastic strain always appears in the solder joint a maximum distance from the centre.   Figure 18-5 illustrates the distribution of von Mises stress in a single ball for various d_b/d_t ratios.

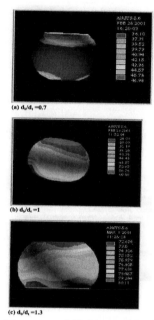

Figure 18-5 Distribution of von Mises stress (MPa) in a solder balls with variable d_b / d_t

    Within a solder ball the maximum stress appears at the interface between solder and silicon chip. It is also noted that the effective stress in the joint increases with the increase in diameter ratio. The distribution of equivalent elastic strain in solder joints is given below:

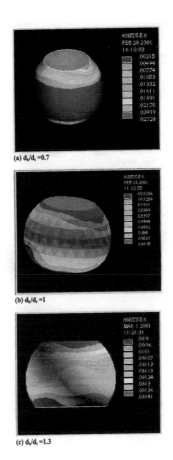

Figure 18-6 Distribution of equivalent elastic strain in solder joints with variable d_b / d_t

From Figure 18-6, the distribution of equivalent inelastic strain in the solder joint can also be seen. The maximum inelastic strain also appears in the solder/silicon chip interface. The solder ball with the largest ratio experiences the highest strain. The estimated values for the three ratios are 2.7%, 3.6% and 8.1%, respectively.

5 FATIGUE LIFE

The Coffin-Manson equation, as revised by Engelmaier, was used to predict the fatigue life of the solder joint. The relationship between fatigue life and the equivalent total strain difference is given by:

$$N_f = \frac{1}{2}\left[\frac{\Delta\gamma}{2\varepsilon_f{}'}\right]^{\frac{1}{c}} \tag{18.5}$$

where, N_f is the mean number of cycles to failure, $\Delta\gamma$ is cyclic shear strain difference and is given by:

$$\Delta\gamma = \frac{\sqrt{3}}{2}\Delta\varepsilon \qquad\qquad .. \tag{18.6}$$

where, $\Delta\varepsilon$ is the total inelastic strain difference in the fourth cycle.
$\varepsilon_f{}'$ is the fatigue ductility coefficient, $\varepsilon_f{}' = 0.325$, c is the fatigue coefficient given by.

$$c = -0.442 - 6\times10^{-4}T_m + 1.74\times10^{-2}\ln(1+f)$$
$$= -0.395 \tag{18.7}$$

where f is the frequency, f=48 cycles/day, T_m is the mean cyclic temperature given by,

$$T_m = 0.5(T_{max} + T_{min}) \tag{18.8}$$

According to the above analysis, the interface between the solder and silicon chip is the sensitive position to failure. The crack will initiate and propagate from here.

The average strain difference between the maximum and minimum during the fourth cycle is obtained from interface elements in the weak joint. The fatigue life is predicted using the Coffin-Manson equation (18.5).

The results, for the three diameter ratios, are given in following Table 18-4.

Table 18-4 Fatigue life of the feeble joints determined from strains estimated in the 4th cycle

d_b / d_t	0.7	1	1.3
Strain difference	0.0273	0.0332	0.045
Fatigue life (cycles)	2205	1343	622

It is observed that the joint with the lower d_b / d_t ratio is the relatively more reliable.

6 CONCLUSIONS

Thermal-mechanical modelling for a solder interconnection in CSP was carried out using a 3D geometry model and a unified viscoplastic (Anand) solder model. The stress-strain distributions in solder joints for a 5x5 solder bump array in CSP with variable solder geometry were obtained. This shows that the maximum equivalent stress and strain occurred in the interface between the silicon chip and solder bumps. The results also indicated that the inelastic strain difference decreases in the solder joint with the decrease of the d_b / d_t ratio, when then solder volume and top pad diameter were kept constant. Employing a modified from of the Coffin-Manson equation, it is possible to predict the fatique life of the joint.

7 REFERENCES

1   XJ Zhao, GQ Zhang, JFJ Caers, LJ Ernst, Finite Element Modeling of Solder Interconnection reliability with Variable Solder geometry, Proceedings of EuroSimE 2001 Benefiting from Thermal and Mechanical Simulation in (Micro)-Electronics, 2001, pp 337-342.
2   R Darveaux, Solder Joint Fatigue Life Model, Proceeding of the TMS, February, 1997, pp 213-218.

INDEX